Estimation Theory in Hydrology and Water Systems

Estimation Theory in Hydrology and Water Systems

K. Nacházel

Faculty of Civil Engineering,
Technical University,
Prague, Czechoslovakia

ELSEVIER

Amsterdam – London – New York – Tokyo 1993

Reviewers:
Doc. Ing. Alexander Puzan, DrSc., Corresponding Member
of the Czechoslovak Academy of Sciences
Prof. Ing. Vojtěch Brožа, DrSc.

Published in co-edition with Academia, Publishing House of the Czechoslovak Academy
of Sciences, Prague

Exclusive sales rights
in the East European Countries, China, Cuba, Mongolia, North Korea and Vietnam
Academia, Publishing House
of the Czechoslovak Academy of Sciences
Prague, Czechoslovakia

in the rest of the world
Elsevier Science Publishers B. V.
Sara Burgerhartstraat 25
P. O. Box 211
1000 AE Amsterdam, The Netherlands

Library of Congress Cataloging-in-Publication Data

Nacházel, Karel.
[Teorie odhadu v hydrologii a ve vodním hospodářství. English]
Estimation theory in hydrology and water systems/K. Nacházel.
p. cm. – (Developments in water science; 42)
Translation of: Teorie odhadu v hydrologii a ve vodním hospodářství.
Includes bibliographical references and index.
ISBN 0–444–98726–6
1. Hydrology–Mathematics. 2. Estimation theory. I. Title. II. Series.
GB656.2.M34N3313 1993
55.48'0151–dc20

ISBN 0–444–98726–6 (vol. 42)
ISBN 0–444–41669–2 (Series)

Printed in Czechoslovakia

OTHER TITLES IN THIS SERIES

1 G. BUGLIARELLO AND F. GUNTER
COMPUTER SYSTEMS AND WATER RESOURCES
2 H. L. GOLTERMAN
PHYSIOLOGICAL LIMNOLOGY
3 Y. Y. HAIMES, W. A. HALL AND H. T. FREEDMAN
MULTIOBJECTIVE OPTIMIZATION IN WATER RESOURCES SYSTEMS:
THE SURROGATE WORTH TRADE-OFF-METHOD
4 J. J. FRIED
GROUNDWATER POLLUTION
5 N. RAJARATNAM
TURBULENT JETS
6 D. STEPHENSON
PIPELINE DESIGN FOR WATER ENGINEERS
7 V. HALEK AND J. ŠVEC
GROUNDWATER HYDRAULICS
8 J. BALEK
HYDROLOGY AND WATER RESOURCES IN TROPICAL AFRICA
9 T. A. McMAHON AND R. G. MEIN
RESERVOIR CAPACITY AND YIELD
10 G. KOVÁCS
SEEPAGE HYDRAULICS
11 W. H. GRAF AND W. C. MORTIMER (EDITORS)
HYDRODYNAMICS OF LAKES: PROCEEDINGS OF A SYMPOSIUM
12–13 OCTOBER 1978, LAUSANNE, SWITZERLAND
12 W. BACK AND D. A. STEPHENSON (EDITORS)
CONTEMPORARY HYDROGEOLOGY: THE GEORGE BURKE MAXEY MEMORIAL
VOLUME
13 M. A. MARINO AND J. N. LUTHIN
SEEPAGE AND GROUNDWATER
14 D. STEPHENSON
STORMWATER HYDROLOGY AND DRAINAGE
15 D. STEPHENSON
PIPELINE DESIGN FOR WATER ENGINEERS
(completely revised edition of Vol. 6 in this series)
16 W. BACK AND R. LETOLLE (EDITORS)
SYMPOSIUM ON GEOCHEMISTRY OF GROUNDWATER
17 A. H. EL-SHAARAWI (EDITOR) IN COLLABORATION WITH S. R. ESTERBY
TIME SERIES METHODS IN HYDROSCIENCES
18 J. BALEK
HYDROLOGY AND WATER RESOURCES IN TROPICAL REGIONS
19 D. STEPHENSON
PIPEFLOW ANALYSIS
20 I. ZAVOIANU
MORPHOMETRY OF DRAINAGE BASINS
21 M. M. A. SHAHIN
HYDROLOGY OF THE NILE BASIN

22 H. C. RIGGS
STREAMFLOW CHARACTERISTICS
23 M. NEGULESCU
MUNICIPAL WASTEWATER TREATMENT
24 L. G. EVERETT
GROUNDWATER MONITORING HANDBOOK FOR COAL AND OIL SHALE DEVELOPMENT
25 W.KINZELBACH
GROUNDWATER MODELLING
26 D. STEPHENSON AND M. E. MEADOWS
KINEMATIC HYDROLOGY AND MODELLING
27 A. M. EL-SHAARAWI AND R. E. KWIATKOWSKI (EDITORS)
STATISTICAL ASPECTS OF WATER QUALITY MONITORING
28 M. JERMAR
WATER RESOURCES AND WATER MANAGEMENT
29 G. W. ANNANDALE
RESERVOIR SEDIMENTATION
30 D. CLARKE
MICROCOMPUTER PROGRAMS IN GROUNDWATER
31 R. H. FRENCH
HYDRAULIC PROCESSES ON ALLUVIAL FANS
32 L. VOTRUBA, Z. KOS, K. NACHÁZEL, A. PATERA AND V. ZEMAN
ANALYSIS OF WATER RESOURCE SYSTEMS
33 L. VOTRUBA AND V. BROŽA
WATER MANAGEMENT IN RESERVOIRS
34 D. STEPHENSON
WATER AND WASTEWATER SYSTEMS ANALYSIS
35 M. A. CELIA ET AL., (EDITORS)
COMPUTATIONAL METHODS IN WATER RESOURCES, 1
MODELING SURFACE AND SUB-SURFACE FLOWS
36 M. A. CELIA ET AL., (EDITORS)
COMPUTATIONAL METHODS IN WATER RESOURCES, 2
NUMERICAL METHODS FOR TRANSPORT AND HYDROLOGICAL PROCESSES
37 D. CLARKE
GROUNDWATER DISCHARGE TEST SIMULATION AND ANALYSIS
38 J. BALEK
GROUNDWATER RESOURCES ASSESSMENT
39 E. CUSTODIO AND A. GURGUI (EDITORS)
GROUNDWATER ECONOMICS
40 D. STEPHENSON
PIPELINE DESIGN FOR WATER ENGINEERS
(third revised and updated edition)
41 D. STEPHENSON AND M. S. PETERSON
WATER RESOURCES DEVELOPMENT IN DEVELOPING COUNTRIES
42 K. NACHÁZEL
ESTIMATION THEORY IN HYDROLOGY AND WATER SYSTEMS

Contents

Preface

Under the contemporary complex hydrological conditions, processing hydrological data in a methodologically correct way has become a task of fundamental importance in the planning of safety, economics and rational utilization of water-engineering projects. The estimation of the representative statistical parameters of the given hydrological series, invariably based upon observation within the limited time interval available, is an essential part of the preparation of these data. And closely linked with this process is the determination of the design quantities and the evaluation of the reliability of the hydrological and water-engineering computations corresponding to the probability character of the original data.

In spite of the fact that probability methods have long been used in Czechoslovakia in processing hydrological data and designing water-engineering projects, the problems of the theory of estimation began to be systematically studied only as late as the seventies. At that time research was stimulated by the current water-engineering practice, in particular the necessity of dealing with the problems of the North Bohemian coalfield. The task involved testing the reliability of the hydrological data and deriving quantities to be applied in anti-flood precautions.

The theory of estimation has recently received much atention in the CIS, the USA, Canada, and in other industrially advanced countries, because the importance of this theory has steadily been rising in various branches of technology. Its rapid development has primarily been facilitated by elaborate simulation models of stochastic processes, as well as by modern computer technology.

The problems of the theory of estimation and the application of this theory to hydrology and to water engineering are both complex and wide ranging, and they have so far not been dealt with systematically in the water-engineering literature.

The present book is thus the first attempt at a complete presentation of these problems. It aims at perfecting the methodology of processing hydrological data water-engineering computations, contributing to the uniform application of scientific methods in practice, and presenting the problems that need to be considered.

The book revises the knowledge in the field of the theory of estimation enriched by the research achievements of the Department of Hydrotechnology of the Faculty of Civil Engineering of the Czech Technical University in Prague, which collaborates closely in this field with the Czech Hydrometeorological Institute in Prague. New knowledge is presented concerning the properties of random and systematic errors in the estimation of parameters of series with various probability distributions; in the field of application of that knowledge the book deals with the estimation of parameters of various types of flow series and with problems of mathematical modelling of these series in view of the bias of the characteristics of a given real sample.

The reader may also be interested in the new developments in the effect of the parameter estimates of flow series on the water-engineering design of reservoirs. In this field the information on the relationship between the theory of estimation and the optimum control of the operation of reservoirs in real time must also be viewed as original.

The book should not be looked upon as a textbook of estimation theory or the theory of reservoir-controlled runoff. It therefore presupposes basic knowledge of probability theory, theory of stochastic processes, mathematical modelling of time series, and computing; and in the field of water-engineering disciplines, hydrology and the theory of runoff control by means of reservoirs. The understanding of the relationships between estimation theory and the optimization of the runoff control by means of reservoirs also requires some knowledge of the elements of the theory of systems.

In writing the book the author has aimed at an objective presentation and clear explanations of the subject matter in order that the results of his research may also be used by practising specialists engaged in the solving of actual water-engineering problems. In this respect the author does not claim to be always mathematically accurate in his explanation of the subject matter: some of the results and arguments arrived at by using the technique of simulation modelling are presented without further proof.

P. Bureš, a mathematics graduate, who worked out the required programmes and supplied the necessary computations, participated in the research over a long period. The author is pleased to acknowledge the helpful collaboration of Doc. Ing. A. Patera, CSc., in the application of the theory of estimation to the designs of reservoirs. The author has greatly profited from, and the level of the book has been enhanced thanks to, the comments of Prof. Ing. V. Broža, DrSc., and Doc. Ing. A. Puzan, DrSc., Corresponding Member of the Czechoslovak Academy of Sciences who revised the work. Mrs A. Kolářová and Mrs M. Pleštilová assisted the author in copying the manuscript and preparing it for press. Lastly, the author wishes to extend his heartfelt thanks to all his collaborators and assistants.

K. Nacházel

Symbols and units

α	(in the theory of reservoir-controled runoff) coefficient of minimum-plus runoff, (in probability theory) density parameter
β	(in the theory of reservoir-controlled runoff) relative magnitude of storage volume
β_v	(in the theory of reservoir-controlled runoff) relative magnitude of the long-term component of storage volume
B	beta function
C_s	coefficient of asymmetry (sample)
C_v	coefficient of variation (sample)
C_s^*	estimate of coefficient of asymmetry
C_v^*	estimate of coefficient of variation
d	probable error
δ	random error
δ_{d0}	Kronecker's delta
Δ	systematic error (in estimation theory), operator (in time series theory)
e	(in probability theory) coefficient of excess, (in estimation theory) efficient estimator
E	mean, average (of a set of random variables or statistical characteristics)
ε_t	random component (white noise)
f	probability density
φ	probability density, (in time series theory) parameter of time series
Φ	distribution function
Φ_i	theoretical quantiles
γ	density parameter

Γ	gamma function
χ	(in sampling theory) random variable with characteristic distribution
J	information (rate of information)
k_i	module coefficient (module)
L	likelihood function
ln	natural logarithm
λ	density parameter
m	density parameter
M	mean, average (of a set of random variables or statistical characteristics)
μ	mean (of a population)
n	sample size (number of members)
ν	number of degrees of freedom or number of realizations
Ω	parametric space
p	(generally as indicator) insurance of water delivery, (in general) probability
p_o	water delivery insurance with respect to repetition
p_t	water delivery insurance with respect to duration
p_d	water delivery insurance (with respect to volume)
P	probability
Ψ	parameter of time series
$r(\tau)$, $r(k)$	standardized autocorrelation function (usually of the sample)
$\varrho(\tau)$, $\varrho(k)$	standardized autocorrelation function (ussually of the population)
s	standard deviation (sample)
S^2	sample variance (unbiassed estimator of σ^2)
σ	standard deviation (of the population)
σ^2	variance (of the population)
t_α	standard random normal variable corresponding to the level of significance $(1 - \alpha)$
T	time or period of repetition
τ	time difference
$\tau(\Theta)$	parametric function
Θ	parameter of probability distribution
u	statistical characteristic of a sample (in general)
u_0	parameter of a population (in general)
w_n	range
x	random variable
x_0	minimum member of a set
z	transformed random variable

Other symbols are explained in the text.

Part I

Foundations of estimation theory

1
Essence of the role of estimation and the fundamental problems of estimation theory

In a number of technological disciplines and in natural sciences we are often faced with the necessity of ascertaining the properties of the random fluctuation of variables, the values of which are obtainable from laboratory or operational experiments, or from the measurement of the respective natural phenomena within a given period of time. The set of a finite number of values obtained in this way can be viewed as a sample of a larger whole, which is then invariably called a population. In practice it is very often ineffective to examine the properties of this set, or these properties can even not be ascertained at all (for instance, if the behaviour of these random variables has been observed within a limited time interval only).

It can easily be shown that a repetition of the experiments or measurement will often supply us with another set of quantities, viz. another sample. The individual samples, although they may be derived from the same population, can thus have different probability properties, which must be examined using statistical methods. In practice, the statistical characteristics of samples, such as means, standard deviations, coefficients of variation, are currently computed, and further characteristics (e. g. sample distribution functions, sample autocorrelation functions) are devised.

Since the properties of the samples differ, the statistical characteristics of the samples, which thus virtually assume the character of random variables, will differ also. We thus get sets of sample characteristics (for instance, sets of sample means, sets of sample standard deviations etc.), for which their own probability properties can again be derived. We therefore speak, for example, of the distribution of sample means, the distribution of sample variances, but also of the mean of sample means, the variance and skewness of sample means, the mean of the sample coefficients of variation etc.

The same analysis can also be worked out for so-called random samples, which are generated from a population in accordance with the rule of randomness. The statistical characteristics, or the probability properties of the whole sets of these characteristics, can thus also be derived for the sets of random samples of the same population.*)

In contrast to characteristics, which are random variables, the probability distribution of a population is described by parameters (e. g. by means, coefficients of variation, coefficients of asymmetry etc.), which are considered to be constants. In computing moments and analyzing them we should therefore strictly distinguish whether we are concerned with moments of samples (characteristics) or moments of a population (parameters).

In practice, it invariably becomes necessary for the unknown parameters of a population to be estimated on the basis of one or several random samples. In statistics [52], such an estimation is understood to be a rule (a decision function) with the help of which the value of an unknown parameter can be estimated on the basis of the probability properties of the sample, either as a number (point estimation) or as an interval within which the unknown parameter is most likely to lie (interval estimation). A detailed analysis of the properties of samples is thus fully justified by the need to estimate the parameters on which further progress in the solution of the problem is very often dependent**).

In this respect, the solution of water-engineering problems can be regarded as a typical case of application of the theory of estimation, because the reliability of the solution (e. g. the reliability of the determination of the design parameters of a water-engineering project) depends fully upon the properties of the hydrological conditions estimated.

---

*) Instead of the older and more descriptive term "sample of the population with distribution $F(x)$" use is now increasingly made in the more recently published literature of a shorter term, viz. "sample of $F(x)$ distribution". In some experiments one sample may be characterized by two, three, or more generally, p vectors of numbers. With each repetition we can thus observe a p-dimensional random vector. We then speak of a random sample of a two- to p-dimensional distribution [65].

**) In their modern conception, the problems of decision-making have gained considerable importance in the decision theory, which is gradually taking shape on the boundaries of the classical disciplines, such as probability theory, logic, psychology, the general theory of management, and cybernetics. These problems are of especial importance in systems disciplines [67], [94]; in water engineering they are particularly applicable to design of water resource systems [115].

The properties of samples and their relationships to the population are dealt with by the so-called random sample theory, which is a branch of mathematical statistics. The methods of parameter estimation themselves belong to the domain of the theory of estimation, which must at present be considered one of the most powerful developments in mathematical statistics.

The complexity and the difficulty of parameter estimation based upon the given characteristics are due to a number of factors, as follows:

1. The distribution of statistical characteristics depends upon the parameters and the type of distribution of the original random variable in the population. With the general approach, the solution of the task will prove extraordinarily difficult, and the literature has so far limited itself invariably to the examination of the properties of the samples of a population with the simplest, i. e. normal (Gaussian) distribution. As far as the more complex types of distribution of a population are concerned, some of the properties of the distribution of characteristics have so far remained virtually unexplained.

2. The relationships between characteristics and parameters are very often rather complex; any analytical expressions facilitating estimation in practice are totally lacking. More significant progress has been made only recently in this field, thanks to the simulation models of stochastic processes, which make it possible for these relationships to be studied with some reliability.

3. The estimation of the type of distribution of the population itself poses particularly great problems. Various methods approach the problem of parameter estimation in a rather simplified way, taking the type of distribution to be known, and only the parameters of that distribution to be unknown. Such assumption can be justified only if there is plenty of experience with the behaviour of the random variables in large populations. It is however generally not possible to determine the type of distribution of the population from a short sample only, and with the behaviour of the variables unknown. In this case it should be admitted that the estimates will range within an interval corresponding to a certain class of the type of distribution. The estimators are then referred to as "robust", and they are expected to exhibit good properties for all the types of distribution considered [37], [64].

4. The estimation of the parameters of higher orders and the more complex (asymmetrical) types of probability distribution is rather complicated due to the effect of both the random errors, and also the systematic errors, which are defined as the difference between the expected values of the set of characteristics and the respective parameter. The properties of systematic errors themselves are relatively complex and they depend upon the length of the given sample, the parameters, and the type of distribution of the population. They have so far been satisfactorily elucidated only for some of the cases of estimation only. Their analysis and the methods of determining them are dealt with in detail in Chapters 4, 5 and 6 of this book.

5. For engineering practice a significant problem is posed by the estimation of parameters in cases where the probability properties of the given series are to be expressed by a larger number of its statistical characteristics. For instance, the average monthly flow series exhibit different properties in different calendar months. Their description therefore requires using at least the first three moments of distribution and, in addition, a system of coefficients of correlation between the flows in the different months. In this way, it therefore becomes necessary to estimate more than fifty parameters per a single profile. It follows that research should not only be oriented towards the elaborate and well-tried methods of estimation, but also towards the production of aids for routine estimation of parameters, and towards the methods of automated computer-aided estimation.

The fundamental problems of the theory of estimation quoted above show that the contemporary methodology of this theory is fairly rich, and that some problems have however so far not been satisfactorily elucidated and solved. Open problems also arise in the field of application of the theory of estimation, viz. utilization of the theory for the estimation of the parameters of the series of concrete types.

2
Development of estimation theory and its application to hydrology and water engineering

2.1 Basic methods of the theory of estimation

The literature [64] shows that the theory of estimation has been developing as part of the theory of probability since the early decades of the nineteenth century. Karl Friedrich Gauss, who formulated in 1821 and 1823 the least squares method, is invariably considered to be the founder of the theory of estimation. But long before Gauss, the theory had already been worked on by Adrien Marie Legendre in 1806.

It was however not until the early years of the twentieth century that the theory of estimation developed more rapidly. Gauss's work was continued by Markov (1900), Aithen (1935), Bose (1950), Rao (1971) and others.

In 1922, R. Fischer, an English statistician, set forth his original ideas in his work on the mathematical foundations of theoretical statistics, pointing to the merits of the method of maximum likelihood [68] as compared with the older moments method. Since then, the maximum likelihood method has attracted the interest of mathematicians, who have been developing it, on the one hand, and on the other, specialists in various disciplines, who have been using it to estimate parameters of the series of various types.

The method has however some drawbacks. In our research [77] we have tested it thoroughly and it has turned out that its application to the engineering practice is rather limited (see Chapter 5, this book).

In Czechoslovakia, the theoretical problems of estimation have been dealt with by Anděl [2, 3], Kubáček [64], Likeš and Machek [65] and others.

The methods of the theory of estimation have been penetrating the field of hydrology and water engineering relatively very slowly. Hydrological data were at first processed using the moments method, though to a limited extent and without the systematic errors of the characteristics being corrected. With the properties of the behaviour of the random and systematic errors unsatisfactorily cleared, the representativeness of the given real sample was at first assumed without any consequences arising from this assumption being considered so far as the reliability of the water-engineering computations is concerned. In Czechoslovakia, comparative analyses of various samples, as well as investigations of the relationships of these samples to the population, were started in the early sixties as part of the development of the application of probability methods to hydraulic engineering.

Since methods of correcting the biassed characteristics of flow series were not available, the same approach was practised in mathematical modelling of the random flow series in the mid-sixties, which flourished particularly owing to the growing need for hydraulic computations of storage reservoirs. Under these circumstances, a substantial merit of designing reservoirs on the basis of long modelled series as compared with the designs based on short real series, was seen in the fact that this method of designing reservoirs gave the expression of the function of reservoirs much higher reliability. The representativeness of the random sequence (in the sense of probability) corresponded, however, to the parent real series.

Neither the Czechoslovak water-engineering literature nor water-engineering practice testify to a spread of the application of the maximum likelihood method, although in 1975 the Czechoslovak National Standard No. 73 6805 [28] recommended the method to be used to estimate the parameters of the more variable flow series. The reason why the maximum likelihood method has so far not enjoyed wider usage is most probably due to the fact that the properties, particularly with the more complex types of probability distribution, are not given satisfactory elucidation; greater attention was thus paid to this method in our research. We compared the properties of the estimators with those of other methods of estimation and were looking for the most adequate ways of determining the required hydrological design quantities.

The quantiles method, set forth by Alekseev [1] in 1960, has a particular relationship to the theory of estimation. The method derives the expressions for the computation of the characteristics of probability distribution from the condition that the theoretical line of transgression determined by these characteristics should cross the empirical quantiles selected. Good approximation is thus achieved of the theoretical to the empirical line of transgression. In some cases, however, the estimates are far from being acceptable, and they can even be worse than in the cases where the moments method is applied [75].

In water engineering, the quantiles method is one of the most frequently used, owing to its simplicity and computational simplicity. It is invariably used for the determination of the design quantities with the lower values of the probability of transgression. For an analysis of the properties of these values from the point of view of the estimation theory, the reader is referred to Chapter 6.

2.2 Methods of examination of the representativeness of sample characteristics based on comparative analysis

For many decades, mathematicians have been interested in the possibility of estimating the unknown representative parameters of a population only on the basis of the time-limited observation of the given variables. The interest in this complex problem has been aroused by awareness of the fact that the probability properties of various samples of the same population can differ substantially – both mutually and from the properties of the population itself. Solving various problems, or drawing conclusions from the observation of a single sample, without the ascertainment of its properties, may thus involve considerable random errors, which can greatly bias the reliability of the solution.

Czechoslovak hydrological and water-engineering computations have always sought to find the representative parameters of flow series that the solution is conditional upon. At the epoch when the properties of the behaviour of the random and the systematic errors had not yet been adequately researched and the methods of estimation elaborated, the assessment of the representativeness of the given real flow series involved the use of various methods of comparative analysis of the properties of the given sample and the properties of the other samples of the same series (under longer observation), or the properties of a related hydrological series (i. e. an analogue, also based on longer observation) and the properties of long geophysical series.

The application of comparative analysis to the investigation of the representativeness of the sample characteristics was closely linked with the development and application of probability methods to water engineering in general. It was in the early sixties that the probability properties of various samples of the same flow series and their non-stationary tendencies and relationship to long-term parameters were examined in Czechoslovakia.

The comparative analysis of the properties of a given series also encompasses the utilization of the correlative relationship of these properties to the respective analogue in parameter estimation. This procedure was of course conditional upon a sufficiently close affinity between the two series compared. It is at present also frequently used for extending shorter series or filling up the missing sections of a series.

The assessment of the representativeness of a given real series and the estimation of the parameters could however be difficult unless an analogue exhibiting a close correlative relationship was found to match that series. And problems were also posed by the representativeness of the distribution of the runoff in the course of the year and the characteristics of the average monthly flow series.

The comparative analysis of the properties of samples of the same series of various length can be viewed as the second methodological trend in the assessment of the representativeness of hydrological series and their statistical characteristics. For instance, several authors focused their attention on the period between 1931 and 1960 (and on the relationship to other series), which for a long time was regarded as a basis for water-engineering analyses. And later, the Hydrological Institute in Prague used the same approach when it compared the representativeness of the flow series in the periods of 1931–1960 and 1931–1970 [20]. The fundamental principle of that method consisted in the characteristics of the flow series in shorter periods being matched against those of a longer series considered to be the basic series. It was the latter series that the relative deviations of the characteristics of the shorter samples were then related to, and the order of the agreement was determined.

The third important method practised in the past in Czechoslovakia so far as the assessment of the representativeness of the designs of reservoirs is concerned, was the comparative analysis of the results of the computation of the storage function of reservoirs in various periods and for various parameters of the runoff control. Much attention was particularly given to the storage function of reservoirs in the 1931–1960 period, and to an analytic comparison with the longer series analogues [116].

2.3 Methods of parameter estimation based upon simulation models of random sequences

The estimation theory aided by the simulation models of random sequences and modern computer technology offers qualitatively new and wider possibilities of estimation of the representative parameters of hydrological series. These methods started to be applied approximately 10 to 15 years ago, and their rapid development was facilitated by the fact that the methodological procedures of modelling random series with the desired probability properties and the techniques of generating random samples from the modelled series had already been fully elaborated.

The advantage of these methods over the preceding partial researches consists above all in their general applicability to the diverse problems of parameter estimation on the basis of short-term observation. Parameters can be estimated for various series (for instance, series of culminating flood flows, series of average annual and monthly flows), and this estimation is not conditional upon a suitable analogue with a close relationship of correlation being available. This book is intended to show that the whole process of estimation can be formalized with the help of algorithms, and computed on powerful computers, which also enable rapid processing of the large number of data supplied by engineering practice.

Whereas in the past the efforts at an exact analytic formulation of the relationships between the parameters of a population and the characteristics of its various samples used to involve mathematical difficulties, the application of the simulation models of random sequences and modern computer technology has removed these difficulties and has led to high reliability, dependent only upon the goodness of fit of the model to the given hydrological conditions, the length of the modelled series and a sufficient number of samples.

The principles of the methods of parameter estimation based upon the simulation models of random sequences, as well as the application of these methods, are dealt with in following chapters. The moments method is considered in Chapter 4, the maximum likelihood method in Chapter 5, and the quantiles method in Chapter 6.

The fundamental problems of the analysis of the time series, the mathematical modelling of these series, and the relationship of this modelling to the theory of estimation are explained in Chapter 7.

3
Sample characteristics. Their distribution

3.1 Definition of characteristics. Their fundamental relationships to parameters

It follows from the preceding chapter that the fundamental properties of samples are described by their characteristics, which are invariably defined as moments of a certain order, or derived from these moments. If the samples stem from the same population, it becomes necessary to derive the properties of the distribution of the whole set of characteristics (moments of the same order) as well as the relationship of the characteristics to parameters.

In a population, the moments are often denoted by Greek letters (e. g. μ, σ etc.), in a sample the notation is by Roman letters (e. g. $\bar{X}$, s etc.).

The expressions for the computation of the characteristics of the discrete sequences that are most frequently applied to the solution of water-engineering problems are given in the following.

The simplest characteristic is the sample mean, which is defined by the following expression:

$$\bar{x} = \frac{1}{n} \sum_{i=1}^{n} x_i \tag{3.1}$$

where $x_1, x_2, \ldots, x_n$ stand for the elements of the sample and n for their number.

The sample range is defined as

$$w_n = x_{\max} - x_{\min}, \tag{3.2}$$

where $x_{\max}$ and $x_{\min}$ denote the maximum and the minimum elements of the given sample.

The sample variance is one of the basic characteristics of the variability (dispersion) of the elements of a given sample. It is defined as the central moment of second order, using the following formula:

$$s^2 = \frac{1}{n} \sum_{i=1}^{n} (x_i - \bar{x})^2 . \tag{3.3}$$

And from this characteristic two more characteristics are derived to express the dispersion: the sample standard deviation as the positive value of the square root of variance, viz.

$$s = + \sqrt{\frac{1}{n} \sum_{i=1}^{n} (x_i - \bar{x})^2} \tag{3.4}$$

and the sample coefficient of variation, which is a dimensionless number:

$$C_v = \frac{s}{\bar{x}} = \sqrt{\frac{1}{n} \sum_{i=1}^{n} \left(\frac{x_1}{x} - 1\right)^2} = \sqrt{\frac{\sum_{i=1}^{n} (k_i - 1)^2}{n}} \tag{3.5}$$

where k_i is the module coefficient.

The asymmetry of the distribution of values x_i round mean $\bar{x}$ is expressed by the coefficient of asymmetry (coefficient of skewness, or skewness), which is given by the following expression:

$$C_s = \frac{1}{ns^3} \sum_{i=1}^{n} (x_i - \bar{x})^3 \tag{3.6}$$

and which is, like the coefficient of variation, a dimensionless number.

The coefficient of excess (also referred to as coefficient of kurtosis, or simply kurtosis) characterizes the accumulation of values x_i in the vicinity of mean $\bar{x}$. It is defined by the following expression:

$$e = \frac{1}{ns^4} \sum_{i=1}^{n} (x_i - \bar{x})^4 - 3 \tag{3.7}$$

which is also a dimensionless number.

From the characteristics computed, the probability properties of the given samples can then easily be inferred. Figure 1 is a visual representation of the effect of the coefficient of asymmetry and the coefficient of excess on the shape of the distribution of the elements of the samples [114, 116].

For the computation of the characteristics with the help of computers the literature offers easily programmable expressions. In computing centres, standard subroutines facilitating the statistical analysis of the sets of data are very often available.

The assessment of the whole set of characteristics derived from the same population involves relatively complex problems. The most significant are:
- the relationship of the characteristics to parameters;

- the effect of the number of elements, n, in a sample (length of sample) on the properties of parameter estimates;
- the distribution of the characteristics.

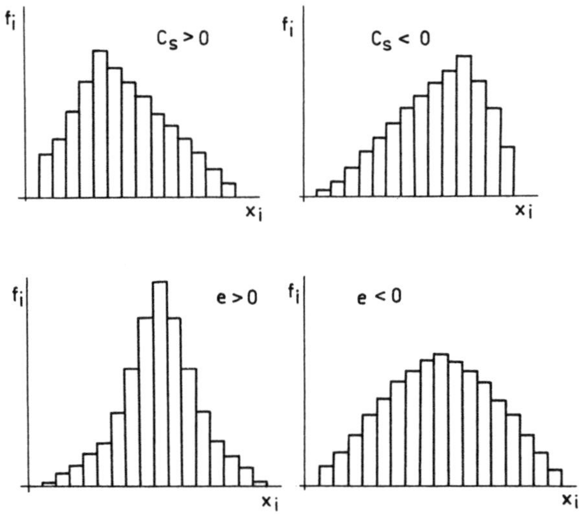

Fig. 1. Distribution of sample values with various extents of skewness and kurtosis.

The first of these problems is most conveniently dealt with by comparing the curve of the characteristics (of the same order) with the respective parameter. If a characteristic is denoted as u, and the respective parameter of the population as u_0, the following relationships arise between the set of characteristics u and u_0:

if for $n \to \infty$ u converges in probability towards parameter u_0, u is called the consistent estimator of variable u_0 [65, 110]. This property of some estimators is given the following written form:

$$\lim_{n \to \infty} P\left(|u - u_0| < \varepsilon\right) = 1 \tag{3.8}$$

for any $\varepsilon > 0$;

if for a given n the expected value of the set of u's equals u_0, viz.

$$E(u) = u_0\,, \tag{3.9}$$

u is called an unbiassed estimator of parameter u_0. It follows that estimator u does not exhibit any systematic error. And with the following inequality:

$$E(u) \neq u_0\,, \tag{3.10}$$

we refer to u as a biassed estimator exhibiting systematic error

$$\Delta = u_0 - E(u)\,. \tag{3.11}$$

Some estimators are interesting owing to the fact that with n increasing, their systematic error Δ decreases boundlessly. In this case we then speak about an asymptotically unbiassed estimator, for which it holds that

$$\lim_{n \to \infty} \Delta = \lim_{n \to \infty} \{u_0 - E(u)\} = 0 . \qquad (3.12)$$

For the schematic diagram of this relationship, see Fig. 2. The curve of the expected values of characteristics $E(u)$ may still be one sidedly biassed below the value of the long-term parameter u_0, but with the length of the sample, n,

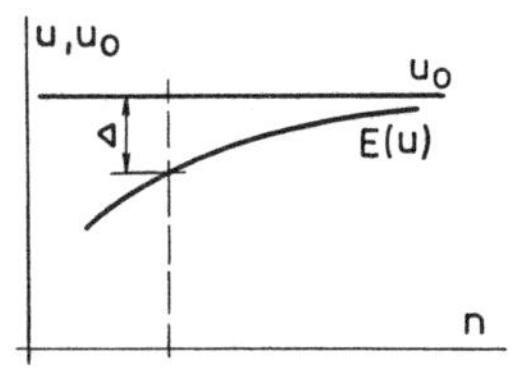

Fig. 2. Schematic diagram of systematic errors.

increasing, it will approximate to that parameter, and the systematic error, Δ, will thus converge towards zero.

The properties of the systematic errors with the individual types of the distribution of a population are dealt with in detail in the following chapters of this book.

When samples are studied, it is essential that evaluation should be undertaken both of the bias of the expected values of the set of characteristics with respect to parameters (i. e. systematic errors), and the bias of the characteristics of the individual samples with respect to parameters. The latter bias is considered to be a random error defined as follows:

$$\delta = u - u_0 . \qquad (3.13)$$

The set of random errors $\delta(u)$ of the same characteristic u is often defined by their variance $\sigma^2(u - u_0)$, which in view of the fact that parameter $u_0 = \text{const.}$ equals

$$\sigma^2(u - u_0) = \sigma^2(u) . \qquad (3.14)$$

In this context, the literature considers estimators u of an unknown parameter u_0 to be the more valuable, the lower is the dispersion defined by equation (3.14). The "best" estimator, often called an efficient estimator, is the one with the lowest dispersion.

No less interesting is the problem of the effect of the number of the elements of a sample, n, in the expressions for the computation of the sample characteristics, on the properties of the unbiassed and best parameter estimators. In the literature, particularly technological literature, we often meet with some difference of opinion concerning the usage of n, i. e. some authors prefer the expression $(n - 1)$, or other values of n's.

Let us clarify the reasons for the preference of these mathematical expressions. The advantage of the relationship incorporating n into the expression for computing sample dispersion (3.3) consists mainly in the fact that it corresponds directly to the definition of the second central sample moment, for which it can be proved [3] that its mean square deviation from parametr σ^2 is less than the mean square deviation of variable

$$S^2 = \frac{1}{n - 1} \sum_{i=1}^{n} (x_i - \bar{x})^2 \tag{3.15}$$

and that it thus holds that

$$E(s^2 - \sigma^2)^2 < E(S^2 - \sigma^2)^2 . \tag{3.16}$$

Relationship (3.16) thus justifies the choice of n from the point of view of the magnitude of the mean square deviation.

From other points of view, however, the coeficient n has a number of disadvantages. The literature dealing with this problem [65] reports that for sample dispersion defined according to (3.15) it holds that

$$E(S^2) = \sigma^2 , \tag{3.17}$$

i. e. the expected value of statistic (3.15) equals variance σ^2 of the population. S^2 is thus an unbiassed estimator of σ^2, which from this point of view justifies the preference for $(n - 1)$ rather than n. In expressions (3.3), (3.4), (3.5) and (3.6), some authors therefore very often substitute $(n - 1)$ for n.

In contrast, the second central sample moment, $M_2 = s^2$, has the following expected value:

$$E(s^2) = \frac{n - 1}{n} \sigma^2 \tag{3.18}$$

so that using coefficient n in variance (3.3) involves a systematic underestimation of the dispersion σ^2.

Anděl [3] draws our attention to yet another important fact, viz. that coefficient $1/n$ in expression (3.3) is also far from being optimal from the point of view of the minimum of the quadratic deviation, and he therefore looks for a number k such that the expression

$$E\left[k \sum_{i=1}^{n} (x_i - \bar{x})^2 - \sigma^2\right]^2$$

is reduced to a minimum value. With

$$Y = \sum_{i=1}^{n} (x_i - \bar{x})^2$$

he arrived at the following relationship:

$$E(kY - \sigma^2)^2 = k^2EY^2 - 2k\sigma^2EY + \sigma^4 =$$

$$= \sigma^4[k^2(n^2 - 1) - 2k(n - 1) + 1] = \sigma^4\left[(n^2 - 1)\left(k - \frac{1}{n + 1}\right)^2 + \frac{2}{n + 1}\right], \tag{3.19}$$

from which it follows that the minimum is reached with $k = 1/(n + 1)$, and that this minimum is equal to $2\sigma^4/(n + 1)$.

The example quoted shows that an unbiassed estimator need in no way be at the same time the best from the point of view of the mean quadratic deviation. Parameter estimation should therefore be judged from several points of view.*)

Even more complex properties are exhibited by the sample coefficients of asymmetry, for which the literature quotes several expressions differing again by coefficient $1/n$ in expression (3.6). The complexity is given by the fact that these coefficients are invariably burdened with considerable random deviations, particularly with shorter samples. However, the expected values of the sample coefficients of asymmetry are also often markedly biassed with respect to the parameters. And, moreover, the numerical procedures for finding the best estimates are very often difficult to carry out.

For the sample coefficient of asymmetry, Czechoslovak researchers currently use the following expression:

$$C_s = \frac{1}{(n - 1)\, s^3} \sum_{i=1}^{n} (x_i - \bar{x})^3 , \tag{3.20}$$

which differs from expression (3.6) only in the substitution of $(n - 1)$ for n. But this modification is not a satisfactory solution for the problem of an unbiassed estimator, which must be determined with the help of more exact methodological procedures based predominantly upon simulation modelling of random sequences.

The problems of the reliability of parameter estimation generally grow with the distributions with a larger number of parameters and with samples of a more limited size. This is because with a larger number of parameters use must be

*) Parameter estimation is thus reminiscent of the multiple-criteria problems of optimization well-known from the systems sciences.

made of the sample moments of higher orders, which are extremely sensitive even to small variations of the individual values, so that for instance one or two inaccuracies of measurement can substantially bias the result of the estimation.

In the literature, the formulation of the role of an estimator and the description of its properties are often very general, use being made of parameter space and parametric functions.

Let us consider a random sample of size n of a distribution that depends upon an uknown parameter $\boldsymbol{\Theta}$. We denote as Ω the set of values that parameter $\boldsymbol{\Theta}$ can acquire, and call this set the parameter space.

The distribution that the random sample is derived from can be a distribution of a single-dimensional random variable, or a distribution of an s-dimensional random vector, $s \geqq 2$. Similarly, $\boldsymbol{\Theta}$ can generally be an r-dimensional vector parameter $\boldsymbol{\Theta} = (\Theta_1, \Theta_2, \ldots, \Theta_r)$, $r \geqq 1$.

From the random sample we need to estimate a certain real function $\tau(\boldsymbol{\Theta}) = \tau(\Theta_1, \Theta_2, \ldots, \Theta_r)$ of the unknown parameter Θ. Function $\tau(\boldsymbol{\Theta})$ is called a parametric function.

The task of estimating function $\tau(\boldsymbol{\Theta})$ involves constructing function $T(\boldsymbol{X})$ on the set of all possible $\boldsymbol{X}$'s such that the distribution of statistic $T = T(\boldsymbol{X})$ will exhibit the closest possible concentration about the correct value of $\tau(\boldsymbol{\Theta})$, with all the values of $\boldsymbol{\Theta}$, if possible. This statistic, $T(\boldsymbol{X})$, is then called the point estimate of function $T(\boldsymbol{\Theta})$.

The estimation of the unknown parameters always involves a certain risk, due to the random character of the sample and the fact that its relationship to function $\tau(\boldsymbol{\Theta})$ is unknown. Incorrect estimation can therefore cause certain losses. In the theory of statistical decision-making these problems are handled with the help of the loss functions. When decisions are made under conditions of uncertainty, the usual requirement is for the mean value of the losses to be as low as possible.

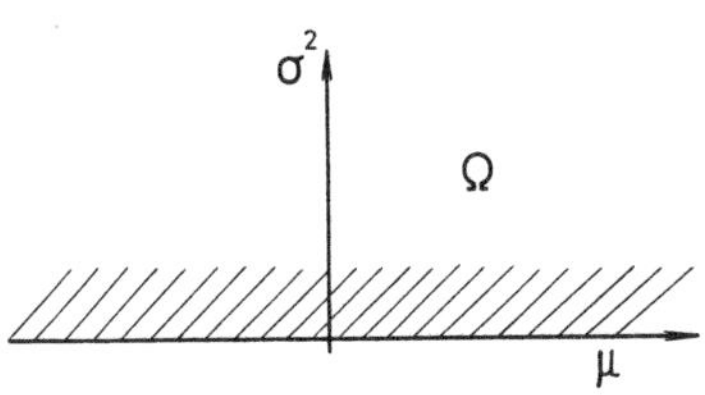

Fig. 3. Example of parameter space.

What parameter space and parametric function are, can be shown on a simple example of normal distribution $N(\mu, \sigma^2)$. The halfplane $-\infty < \mu < \infty$, $\sigma^2 > 0$ (see Fig. 3) is the parameter space, and $\tau(\mu, \sigma^2) = \mu$ (mean value of distribution), $\tau(\mu, \sigma^2) = \sigma^2$ (distribution variance), $\tau(\mu, \sigma^2) = \mu + t_p\sigma$ (distribution quantile) etc. can for example be parametric functions.

The $T(\mathbf{X})$ estimator is regarded as an unbiassed estimator of the parametric function $\tau(\boldsymbol{\Theta})$, provided it holds according to (3.9) that

$$E[T(\mathbf{X})] = \tau(\boldsymbol{\Theta}) \tag{3.21}$$

for all $\boldsymbol{\Theta} \in \Omega$.

The difference

$$\Delta(\boldsymbol{\Theta}) = E[T(\mathbf{X})] - \tau(\boldsymbol{\Theta}) \tag{3.22}$$

is then referred to as a biassed estimator.

The $T(\mathbf{X})$ estimator is the best unbiassed estimator of function $\tau(\boldsymbol{\Theta})$ if

a) the $T(\mathbf{X})$ estimator is unbiassed,

b) for any other unbiassed estimator $T'(\mathbf{X})$ it holds that

$$\sigma^2[T(\mathbf{X})] \leqq \sigma^2[T'(\mathbf{X})]\,. \tag{3.23}$$

The best unbiassed estimator often proves to be an acceptable tool for the tasks of estimation. In some cases, however, an unbiassed estimator may not exist at all, or its construction may be too difficult or even completely unknown. It then becomes necessary that both the variance and the bias of the estimator should be subjected to assessment. The so-called mean square error (deviation) of the estimator, which is defined as variable

$$K(\boldsymbol{\Theta}) = E\{[T(\mathbf{X}) - \tau(\boldsymbol{\Theta})]^2\} = \Delta^2(\boldsymbol{\Theta}) + \operatorname{var}\,[T(\mathbf{X})] \tag{3.24}$$

is an important criterion.

Sometimes it is required to assess the quality of the estimator only according to the asymptotic properties. The usual requirement is that the estimator should be consistent, i. e. with the number of observations, increasing, the estimate will converge towards the actual value of function $\tau(\Theta)$. Property (3.8) can thus be written in a more general form as

$$\lim P(|T(\mathbf{X}) - \tau(\boldsymbol{\Theta})| < \varepsilon) = 1 \tag{3.25}$$

for any $\varepsilon < 0$ and for all $\boldsymbol{\Theta} \in \Omega$ (i. e. the so-called convergence in probability).

Sometimes we must accept an estimator the bias of which declines only with increasing number of observations. In this case we speak about an asymptotically unbiassed estimator, for which it generally holds that

$$\lim_{n\to\infty} \Delta(\boldsymbol{\Theta}) = \lim_{n\to\infty} \{E[T(\mathbf{X})] - \tau(\boldsymbol{\Theta})\} = 0\,. \tag{3.26}$$

The condition of asymptotic unbiassedness, together with the condition

$$\lim_{n\to\infty} \operatorname{var}\,[T(\mathbf{X})] = 0\,, \tag{3.27}$$

are regarded as satisfactory as far as the consistency of the estimator is concerned.

A consistent estimator can be exemplified by the estimation of variance σ^2 of the distribution with the final fourth central moment μ_4. Let $\mathbf{X} = (x_1, x_2, \ldots, x_n)$ be a random sample, and let us consider the variance σ^2 estimators in the following form:

$$T_1 = \frac{1}{n-1} \sum_{i=1}^{n} (x_i - \bar{x})^2 = S^2\,,$$

$$T_2 = \frac{1}{n} \sum_{i=1}^{n} (x_i - \bar{x})^2 = \frac{n-1}{n} S^2 = M_2\,.$$

It can be shown that the variance of statistic S^2 is given by:

$$\operatorname{var}(S^2) = \frac{\mu_4}{n} - \frac{n-3}{n(n-1)}\sigma^4, \qquad n \geqq 3\,, \tag{3.28}$$

where μ_4 is the fourth central moment of the distribution that the sample is derived from.

The following asymptotic relationship therefore holds:

$$\lim_{n\to\infty} \operatorname{var}(T_1) = 0\,.$$

T_1 is thus a consistent estimator of σ^2.

Statistic T_2 is an asymptotically unbiassed estimator of σ^2, and for its variance, $\operatorname{var}(T_2)$, it holds that

$$\lim_{n\to\infty} \operatorname{var}(T_2) = \lim_{n\to\infty} \operatorname{var}(T_1) = 0\,,$$

so that T_2 is also a consistent estimator of σ^2.

For the construction of the best unbiassed estimators a special class of distribution – the so-called exponential class of distribution – is of importance. Variable X has a distribution of an exponential type, if its probability density function $f(x)$ can be written in the following form [35, 65, 92]:

$$f(x; \boldsymbol{\Theta}) = \exp\left[\sum_{j=1}^{k} Q_j(\boldsymbol{\Theta})U_j(x) + R(\boldsymbol{\Theta}) + V(x)\right] \tag{3.29}$$

and if it satisfies the following conditions:

set $\{x \mid f(x; \boldsymbol{\Theta}) > 0\}$ is independent of $\boldsymbol{\Theta}$, (3.30)

parameter space Ω contains a k-dimensional interval, i. e. points $\boldsymbol{\Theta}$ for which $f(x; \boldsymbol{\Theta})$ is the probability density function. (3.31)

As an example of a distribution belonging to the exponential class, let us quote the log-normal distribution [35]. Its density

$$f(x;\, \mu, \sigma^2) = \frac{1}{\sigma x \sqrt{2\pi}} \exp\left[-\frac{1}{2\sigma^2} (\ln x - \mu)^2\right], \qquad x > 0$$

can be written in the following form:

$$f(x;\, \mu, \sigma^2) = \exp\left[-\frac{1}{2\sigma^2} (\ln x)^2 + \frac{\mu}{\sigma^2} \ln x - \frac{\mu^2}{2\sigma^2} - \frac{1}{2} \ln \sigma^2 - \ln x - \right.$$

$$\left. - \ln \sqrt{2\pi}\right], \qquad x > 0\,,$$

i. e. in the form of (3.29), where

$$k = 2,\, \boldsymbol{\Theta} = (\mu, \sigma^2)\,, \qquad Q_1(\boldsymbol{\Theta}) = -\frac{1}{2\sigma^2}, \qquad Q_2(\boldsymbol{\Theta}) = \frac{\mu}{\sigma^2},$$

$$U_1 = (\ln x)^2\,, \qquad U_2 = \ln x, \qquad R(\boldsymbol{\Theta}) = -\frac{1}{2}\left(\frac{\mu^2}{\sigma^2} + \ln \sigma^2\right) - \ln \sqrt{2\pi}\,,$$

$$V(x) = -\ln x\,.$$

Parameter space $\Omega = \{(\mu, \sigma^2) \mid -\infty < \mu < \infty\,,\ \sigma^2 > 0\}$ is a half-plane; set $\{x | f(x;\, \mu, \sigma^2) > 0\} = (0, \infty)$ is thus independent of (μ, σ^2).

But, for instance, uniform distribution within the interval $(0, \Theta)$ does not belong to the exponential class, because its density equals

$$f(x;\, \Theta) = \frac{1}{\Theta}, \qquad 0 < x < \Theta\,,$$

so that the set $\{x \mid f(x;\, \Theta) > 0\}$ depends on Θ, and it does not satisfy condition (3.30).

Special statistical literature [36, 65] shows that the exponential class of probability distribution is of considerable practical importance, particularly as far as the formulation of the best unbiassed estimators is concerned. These estimators are often sought with the help of the so-called sufficient statistics.

Sufficient statistics can be defined with the help of the joint density of a random vector from the distribution of the exponential type. Joint density is thus resolved into several functions that depend both upon value x of the random variable and upon the value of the unknown parameter Θ.

If $\boldsymbol{X} = (X_1, \ldots, X_n)$ is a random sample of the distribution of the exponential type, then the joint density of random vector $\boldsymbol{X}$ equals

$$f(x; \boldsymbol{\Theta}) = \exp\left[\sum_{j=1}^{k} Q_j(\boldsymbol{\Theta}) \sum_{i=1}^{n} U_j(x_i) + nR(\boldsymbol{\Theta}) + \sum_{i=1}^{n} V(x_i)\right] =$$

$$= \exp\left[\sum_{j=1}^{k} Q_j\,(\boldsymbol{\Theta}) S_j(\boldsymbol{x}) + nR(\boldsymbol{\Theta}) + V(x)\right] \tag{3.32}$$

where

$$S_j(\boldsymbol{x}) = S_j(x_1, \ldots, x_n) = \sum_{i=1}^{n} U_j(x_i), \qquad j = 1, \ldots, k,$$

$$V(x) = \sum_{i=1}^{n} V(x_i)\,. \tag{3.33}$$

Statistics $S_j(\boldsymbol{X}), \ldots, S_k(\boldsymbol{X})$, given by expressions (3.33), represent the highest possible reduction of the results of observation, and the most expedient replacement of all the n observations by a lower number of data. They are therefore referred to as minimum sufficient statistics. The estimators with the best properties for functions $\tau(\boldsymbol{\Theta})$ of the parameters of the distribution of the exponential class are invariably functions of statistics.

It can be shown [35] that, for instance, statistic $\sum_{i=1}^{n} x_i = n\bar{x}$ is a sufficient statistic for parameter λ of Poisson's distribution, $Po(\lambda)$, for parameter μ of the Gaussian distribution $N(\mu, \sigma^2)$ with σ^2 known, and for parameter δ of the exponential distribution $E(0, \delta)$. And $\sum_{i=1}^{n} (x_i - \mu)^2$ is a sufficient statistic for parameter σ^2 of distribution $N(\mu, \sigma^2)$ with μ known.

In the assessment of variance the concept of the so-called information is of particular importance. With, for example, two statistics with the same expected values, the statistic with lower variance is always considered to be the better unbiassed estimator of parametric function $\tau(\Theta)$. In this context, we are of course interested in whether it is possible to ascertain the lower limit of the variance of the unbiassed estimators of the parametric function $\tau(\Theta)$.

Let us suppose that the distribution of a random variable has density $f(x; \Theta)$ dependent upon parameter Θ (for simplicity, a one-dimensional parameter), drawing values from an open interval Ω on straight line. Let $f(x;\Theta)$ satisfy the following conditions:

$$M = \{x \mid f(x; \Theta) > 0\rangle \text{ independent of } \Theta\,, \tag{3.34}$$

$$\int_M \frac{f(x; \Theta)}{\partial \Theta}\,\mathrm{d}x = \int_M \frac{\partial \ln f(x; \Theta)}{\partial \Theta} f(x; \Theta)\,\mathrm{d}x = 0 \tag{3.35}$$

for all $\Theta \in \Omega$;

$$J(\Theta) = \int_M \left[\frac{\partial \ln f(x; \Theta)}{\partial \Theta}\right]^2 f(x; \Theta)\,\mathrm{d}x \tag{3.36}$$

is a finite positive number for every $\Theta \in \Omega$. The systems of densities $\{f(x; \Theta), \Theta \in \Omega\}$ satisfying the conditions quoted above are considered to be regular. Function $J(\Theta)$ of parameter Θ is then called information (Fisher's measure of information) pertinent to $f(x; \Theta)$. The derivative of the natural logarithm of function (3.29) with respect to Θ is obviously equal to

$$\sum_{j=1}^{k} Q'(\Theta)\, U_j(x) + R'(\Theta)\,.$$

Information $J(\Theta)$ can then be derived from equation (3.29) of probability density function $f(x; \Theta)$ as the variance of the derivative of its natural logarithm with respect to Θ, i. e. in the following form:

$$J(\Theta) = E\left\{\left[\sum_{j=1}^{k} Q_j'(\Theta)\, U_j(x) + R'(\Theta)\right]^2\right\}. \tag{3.37}$$

If the second derivatives $Q''(\Theta)$ and $R''(\Theta)$ with respect to Θ exist, $J(\Theta)$ can be expressed in the following form [35]:

$$J(\Theta) = -Q''(\Theta)\, E[U(x)] - R''(\Theta) \tag{3.38}$$

for one-dimensional parameter Θ and, analogously, also for a multi-dimensional parameter.

Information $J(\Theta)$ is made use of in the Rao–Cramer theorem, which is of fundamental importance in this field as far as the examination of the lower limit of the mean quadratic error, $R(T - \Theta)^2$, of estimator T, and the question of when that limit is reached [3, 35, 92], are concerned.

Let T be an estimator of Θ such that $ET^2 > \infty$ holds for every $\Theta \in \Omega$. Let $\Delta(\Theta) = ET - \Theta$ be the bias of estimator T. Let us further assume that the following conditions are satisfied:

a) the system of densities $f(x; \Theta)$ is regular,
b) derivative $\Delta'(\Theta)$ exists at every point $\Theta \in \Omega$,
c) it holds that

$$\frac{\partial}{\partial \Theta} \int_M T(x) f(x; \Theta) \, d\mu(x) = \int_M T(x) f'(x; \Theta) \, d\mu(x) . \tag{3.39}$$

For every $\Theta \in \Omega$ it then holds that

$$E(T - \Theta)^2 \geqq \frac{[1 + \Delta'(\Theta)]^2}{J(\Theta)} . \tag{3.40}$$

The estimator T satisfying the conditions of the Rao–Cramer theorem is called regular. For the unbiassed regular estimator it holds that

$$\text{var } T \geqq \frac{1}{J(\Theta)} . \tag{3.41}$$

The number $1/J(\Theta)$ is referred to as Rao–Cramer's lower limit of the variance of the unbiassed regular estimator. This theorem thus gives accurate expression to the intuitively felt fact that the accuracy of the estimator cannot arbitrarily be enhanced. In practice, this limit is often merely an unattainable ideal, which should of course be approximated to as close as possible. In this respect, the concept of efficiency, i. e. relative accuracy of the estimator with respect to the most accurate estimator possible, proves to be a suitable criterion of accuracy. The efficiency of an unbiassed regular estimator is defined as

$$e = \frac{1}{J(\Theta) \text{ var } T} . \tag{3.42}$$

It thus obviously holds that

$$0 \leqq e \leqq 1 . \tag{3.43}$$

With $e = 1$, the estimator is called efficient. As "efficient" in this sense we thus regard an unbiassed regular estimator the variance of which, var T, equals the lower limit of variances $1/J(\Theta)$.

Example [35]

Normal distribution $N(\mu, \sigma^2)$ with parameter σ^2 known is a distribution of the exponential type, which can be expressed in the following form:

$$f(x; \mu) = \exp \{x\mu/\sigma^2 - \mu^2/2\sigma^2 - (1/2) \ln (2\pi\sigma^2) - x^2/2\sigma^2\} ,$$
$$-\infty < x < \infty . \tag{3.44}$$

The interval $(-\infty, \infty)$, within which $f(x; \mu) > 0$, is independent of μ. In expression (3.44) the individual terms in the exponent have the following meaning:

$$\begin{aligned} U(x) &= x\,, \\ Q(\mu) &= \mu/\sigma^2\,, \\ R(\mu) &= -\mu^2/2\sigma^2 - (1/2)\ln(2\pi\sigma^2), \\ V(x) &= -x^2/2\sigma^2\,, \end{aligned}$$

so that

$$Q'(\mu) = 0, \qquad R''(\mu) = -1/\sigma^2\,.$$

As regards information, the following relationship thus holds according to (3.38):

$$J(\mu) = \frac{1}{\sigma^2}\,. \tag{3.45}$$

And similarly, for distribution $N(\mu, \sigma^2)$ with the expected value of μ known, the following relationship can be derived:

$$J(\sigma^2) = \frac{1}{2\sigma^4}\,. \tag{3.46}$$

3.2 Problems of the distribution of characteristics

Finding the probability distribution of the individual sample characteristics derived from one and the same population is a most difficult task. For the random samples of normal distribution the literature quotes analytical expressions of the distribution of their characteristics. In the more complex cases it becomes necessary to apply modelling procedures.

From among the so-called sampling distributions (the term being derived from the fact that they are concerned with probability distributions of sample characteristics) the most frequent use is made of distribution t, distribution χ^2, and distribution F.

For a universe with normal distribution it can be shown that the sample means, $\bar{x}$, also exhibit Gaussian distribution. The mean of the sample means equals the mean of the universe, viz.

$$E(\bar{x}) = \mu\,. \tag{3.47}$$

The variance of the sample means, $\sigma^2(\bar{x})$, is n-times less than the variance of the universe, σ^2,

$$\sigma^2(\bar{x}) = \frac{\sigma^2}{n}. \tag{3.48}$$

For the standard deviation of the sample means it thus holds that

$$\sigma(\bar{x}) = \frac{\sigma}{\sqrt{n}}. \tag{3.49}$$

And if random variable

$$t' = \frac{\bar{x} - \mu}{\sigma(\bar{x})} = \frac{(\bar{x} - \mu)\sqrt{n}}{\sigma} \tag{3.50}$$

is introduced, it becomes evident that

$$E(t') = 0, \qquad \sigma(t') = 1,$$

and that the random variable t' also has Gaussian distribution.

If another random variable is introduced,

$$t = \frac{(\bar{x} - \mu)\sqrt{n}}{s}, \tag{3.51}$$

which differs from t' by also having a random variable, s, in the denominator, it can be shown [110] that variable t exhibits the Student distribution of probability with $k = n - 1$ degrees of freedom.

The properties of t-distribution (Student distribution) have been described in detail [65, 110, 114); they are therefore not subjected to any particular analysis in this book. Probability density $\varphi(t)$ is a bell-shaped symmetrical curve exhibiting higher standard deviation and greater kurtosis than the Gaussian distribution. With the number of the degrees of freedom, k, increasing, $\varphi(t)$ will approximate to standardized normal distribution.

The distribution of the sample means of a population not exhibiting normal distribution is much more complex. With the length of the sample, n, increasing, it will some times approximate to normal distribution.

For the distribution of sample variances s^2 derived from a population with Gaussian distribution and variance σ^2, both the probability density [114]

$$\varphi(s^2) = \frac{1}{\Gamma\left(\frac{n-1}{2}\right)} \left(\frac{n}{2\sigma^2}\right)^{(n-1)/2} (s^2)^{(n-3)/2} \exp\left\{-\frac{n}{\sigma^2} \cdot \frac{s^2}{2}\right\} \tag{3.52}$$

and the distribution function

$$\Phi(s^2) = \frac{1}{\Gamma\left(\frac{n-1}{2}\right)} \left(\frac{n}{2\sigma^2}\right)^{(n-1)/2} \exp\left\{-\frac{n}{2\sigma^2}\int_0^{s^2} (s^2)^{(n-3)/2}\right\} \exp\{-s^2\} \, ds^2 \quad (3.53)$$

can readily be derived.

This is an asymmetrical distribution within the domain of $\langle 0; \infty)$; with $n \geqq 4$, function $\varphi(s^2)$ is a bell-shaped curve, which will become more symmetrical, and will approximate to Gaussian distribution with n increasing.

The mean of sample variances, $E(s^2)$, is given by expression (3.18), and for the variance of variances it holds that

$$\sigma^2(s^2) = \frac{2(n-1)}{n^2}\sigma^4, \quad (3.54)$$

so that the standard deviation of the sample variances equals

$$\sigma(s^2) = \frac{\sqrt{2(n-1)}}{n}\sigma^2. \quad (3.55)$$

The transformation

$$\chi^2 = n\frac{s^2}{\sigma^2} \quad (3.56)$$

converts the distribution of the sample variances to distribution χ^2, which exhibits $\nu = n - 1$ degrees of freedom. With the number of the degrees of

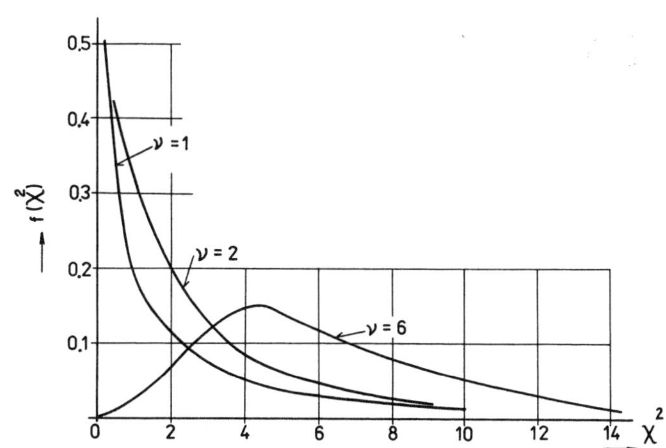

Fig. 4. Distribution χ^2.

freedom growing, distribution χ^2 will approximate to Gaussian distribution (Fig. 4). Ever since distribution χ^2 has been tabulated, its practical applicated has widened.

The probability density of variable χ^2 is equal to

$$\varphi(\chi^2) = \frac{1}{2^{\nu/2}\,\Gamma\left(\frac{\nu}{2}\right)} (\chi^2)^{(\nu/2)-1} \exp\left\{-\frac{1}{2}\chi^2\right\} \tag{3.57}$$

and the distribution function is given by the following relationship:

$$\Phi(\chi^2) = \frac{1}{2^{(\nu/2)}\,\Gamma\left(\frac{\nu}{2}\right)} \int_0^{\chi^2} (\chi^2)^{(\nu/2)-1} \exp\left\{-\frac{1}{2}\chi^2\right\} d\chi^2 \tag{3.58}$$

where ν stands for the number of the degrees of freedom.

And analogously with transformation (3.56) it can be shown [110] that variable

$$\chi = \sqrt{n}\,.\,\frac{s}{\sigma} \tag{3.59}$$

has distribution χ with $\nu = n - 1$ degrees of freedom, which proves to be suitable for the examination of the distribution of sample standard deviation s.

Distribution F (Snedecorian, also Fisher–Snedecorian) is manifested by random variable F defined as the ratio of two mutually independent random quantities with distributions χ_1^2, χ_2^2, and degrees of freedom ν_1 and ν_2:

$$F = \frac{\chi_1^2/\nu_1}{\chi_2^2/\nu_2}\,. \tag{3.60}$$

The probability density, $\varphi(F)$, and the distribution function of the variable, $\Phi(F)$, are expressed as follows:

$$\varphi(F) = \left(\frac{\nu_1}{\nu_2}\right)^{(\nu_1/2)} \frac{1}{\mathrm{B}\left(\frac{\nu_1}{2}, \frac{\nu_2}{2}\right)} F^{(\nu_1/2)-1} \left(1 + \frac{\nu_1}{\nu_2}F\right)^{-(\nu_1+\nu_2)/2}, \tag{3.61}$$

$$\Phi(F) = \left(\frac{\nu_1}{\nu_2}\right)^{(\nu_1/2)} \frac{1}{\mathrm{B}\left(\frac{\nu_1}{2}, \frac{\nu_2}{2}\right)} \int_0^F F^{(\nu_1/2)-1} \left(1 + \frac{\nu_1}{\nu_2}F\right)^{-(\nu_1+\nu_2)/2} dF \tag{3.62}$$

where B denotes the beta function.

Distribution F is asymmetric (Fig. 5), and with the values of ν_1 and ν_2 increasing, it will gradually approximate to Gaussian distribution. If only one of parameters ν_1, ν_2 increases, distribution F will approximate to distribution χ^2.

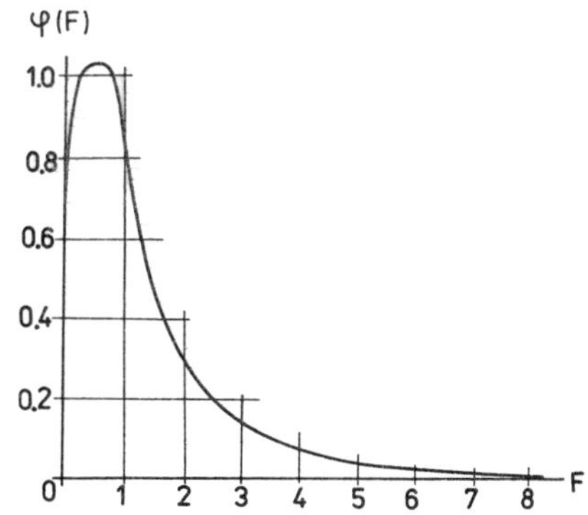

Fig. 5. Distribution F for $\nu_1 = 4$ and $\nu_2 = 3$.

With $\nu_1 = 1$ and $\nu_2 \to \infty$ the distribution of quantity F will approximate to distribution t.

Distribution F is often used for testing the difference between the variances of two random samples derived from populations exhibiting the same variance σ^2. In these tests, use is increasingly made of tabulated critical values of F_p at a certain level of significance p.

A survey of the knowledge of the behaviour of the characteristics and their distribution gained so far, shows that the relationships between the characteristics and the unknown parameters have as yet been reliably formulated only for a population exhibiting Gaussian distribution. So far as populations not exhibiting normal distribution are concerned, these relationships are much more complex. It can be shown that in such cases we can, with some approximation, assume Gaussian distribution only with the sample means (viz. with longer samples). With higher moment characteristics this approximation is inadmissible, which thus makes it necessary to seek the methodological procedures that could help to define these relationships.

Relatively great attention has been given to these problems in the Soviet water-engineering literature ([96] etc.), in which empirical formulae are derived for standard deviations of the sample characteristics of flow series with asymmetrical Pearsonean distribution. We tested the reliability of these formulae using the simulation models of random sequences. (For the results of these tests the reader is referred to Section 4.3).

3.3 Estimators of autocorrelation function and spectral density. Problems of filtration

Apart from the moments of distribution, the significant characteristics of samples also include the autocorrelation function and the periodogram. These characteristics find wide application in such technological disciplines where the solution of problems depends upon information concerning the properties of the internal structure of the samples (for instance, on the tendency in the chronological arrangement of the values of the elements of discrete sequences). In hydrology and in water engineering they have already also become indispensable. The autocorrelation function proves to be indispensable in the examination of the properties of hydrological series and in mathematical modelling of these series; and the computation of the capacity of storage reservoirs is to a great extent dependent upon the calculation of the autocorrelation function. The spectral analysis of hydrological series serves as a basis for the construction of the periodic models of these series, or for the estimation of the future elements of a series.

The correlation and the spectral analyses of time series are at present dealt with in detail by the theory of random processes, which examines the properties of these series using elaborate methodological procedures. Despite these advances, the important problem of the estimation of the correlation function or spectral density on the basis of a single real sequence of finite length has so far remained to a great extent uninvestigated. The examination of the properties of these estimators is of course a rather complex problem, the solution of which depends upon the probability properties of both the original data and the universe. (These problems are dealt with in more detail in Section 10.2).

The standardized sample autocorrelation function is invariably defined as follows:

$$r(\tau) = \frac{\sum_{i=1}^{n-\tau} (x_i - \bar{x}_i)(x_{i+\tau} - \bar{x}_{i+\tau})}{s_i\, s_{i+\tau}(n - \tau - 1)} \tag{3.63}$$

where n stands for the length of the sample (realization of the sequence), and $\bar{x}_i$, $\bar{x}_{i+\tau}$ for the expected values of random variables x_i and $x_{i+\tau}$.

The reliability limits (confidence zone) are determined by the following formula:

$$r_{t_\alpha}(\tau) = \frac{-1 \pm t_\alpha \sqrt{n - \tau - 2}}{n - \tau - 1} \tag{3.64}$$

where t_α is the standard random normal variable corresponding to the level of significance $(1 - \alpha)$.

Anděl [2] shows that the correlation function can be estimated under certain assumptions concerning the properties of the random process or sequence, among which belong above all the stationarity and the ergodicity of the process or sequence.

In his theory of stationary random functions, Jaglom [39] discusses in detail the assumptions mentioned above, as well as their considerable practical importance for the estimation of the correlation function on the basis of a single real process. If the following relationships hold for the unstandardized correlation function $R(\tau)$,

$$\lim_{T\to\infty} \frac{1}{T} \int_0^T R(\tau)\, d\tau = 0\,, \tag{3.65}$$

or

$$\lim_{n\to\infty} \frac{1}{n+1} \sum_{\tau=0}^{n} R(\tau) = 0\,, \tag{3.66}$$

then the expected value μ and the autocorrelation function $R(\tau)$ of a stationary random process can, with some approximation, be computed from the following formulae:

$$\mu \approx \frac{1}{n} \sum_{k=1}^{n} x^{(1)}(k\Delta)\,, \tag{3.67}$$

$$R(\tau) \approx \frac{1}{n} \sum_{k=1}^{n} x^{(1)}(k\Delta + \tau)\, x^{(1)}(k\Delta)\,, \tag{3.68}$$

where Δ denotes a short time interval, n is selected so that $n\Delta = T$ may be great enough, and $x^{(1)}$ stands for the elements of the given realization. And analogously with equations (3.67) and (3.68) Jaglom estimates the expected value and the autocorrelation function of a random sequence using the following expressions:

$$\mu \approx \frac{1}{n+1} \sum_{t=0}^{n} x^{(1)}(t)\,, \tag{3.69}$$

$$R(\tau) \approx \frac{1}{n+1} \sum_{t=0}^{n} x^{(1)}(t + \tau)\, x^{(1)}(t)\,, \tag{3.70}$$

where $x^{(1)}(t)$ again stand for the values of the elements of the realization observed.

Let us recall that the asymptotic relationships (3.65) and (3.66) very often hold in practice if the coefficients of correlation, $R(\tau)$, converge to zero for $\tau \to \infty$, i. e. if the relationships of correlation between the variables grow boundlessly weaker with increasing time remoteness τ.

With the longer real or synthetic series it can easily be demonstrated that the autocorrelation functions of the individual samples, though they may be derived from one and the same series (i. e. the same population), can differ quite considerably. That is why the study of the behaviour of the autocorrelation functions is of immense importance, for it provides the basis for the decision on the most suitable type of model for a given series. For instance, the application of the Box–Jenkins methodology [14] often involves determining the value $\tau = \tau_0$, beyond which the autocorrelation function will equal zero, or ascertaining whether such a value τ_0 exists at all. For example, for the model of the following form,

$$Q_t = \varepsilon_t + \psi_1 \varepsilon_{t-1}, \tag{3.71}$$

where ε_t denotes white noise, and ψ_1 a parameter, it holds for the first autocorrelation coefficient [2] that

$$\varrho(1) = \frac{1}{1 + \psi_1^2}, \tag{3.72}$$

$$\varrho(\tau) = 0 \quad \text{for} \quad \tau > 1, \tag{3.73}$$

so that in this case $\tau_0 = 1$.

But the cause of the greatest difficulties as far as the selection of a convenient type of model is concerned, is the fact that the autocorrelation function $\varrho(\tau)$ pertinent to the population is actually unknown. It thus becomes essential that an assessment should be undertaken of how reliably the estimated sample autocorrelation function $r(\tau)$ will substitute for it. In this context, attention should also be given to the admissible range of variation of the $r(\tau)$ values about zero, for which it can, with a priori given reliability, be assumed that $\varrho(\tau) = 0$.

Use can here be made of the standard deviation of estimator $r(\tau)$ of the autocorrelation function $\varrho(\tau)$. If $\varrho(\tau) = 0$ for $\tau > \tau_0$, then according to Bartlett's approximation [8], with the process normal, it holds that

$$\sigma[r(\tau)] \cong \sqrt{\frac{1}{n}\left(1 + 2\sum_{j=1}^{\tau_0} r^2(j)\right)}, \qquad \tau > \tau_0. \tag{3.74}$$

For the decision on whether $\varrho(\tau) = 0$ is to be adopted, the $|r(\tau)|$ value must be compared with the value of $2\sigma[r(\tau)]$. Use will also have to be made of the fact that the normal random variable with zero expected value will exceed in absolute value the double of its standard deviation with an approximate probability of only 5 percent.

Particularly difficult is the estimation of spectral density linked with the autocorrelation function by means of the Fourier transformation (e. g. [2, 26, 53, 111]), so that one statistical characteristic can easily be converted to another, and vice versa.

In statistical literature particular attention is paid to the problem of the periodicity of real sequences of finite lengths, and to asymptotic relationships with $n \to \infty$.

Here, statistical analysis is based on the so-called periodogram, which is defined by the following formula for the finite sequence of random variables $x_1, x_2, \ldots, x_n$:

$$I_n(\lambda) = \frac{1}{2\pi n}\left|\sum_{t=1}^{n} x_t\, \mathrm{e}^{-it\lambda}\right|^2, \qquad -\pi \leqq \lambda \leqq \pi. \tag{3.75}$$

This formula can also be written:

$$I_n(\lambda) = \frac{1}{2\pi n}\left|\sum_{t=1}^{n} x_t \cos t\lambda - \mathrm{i}\sum_{t=1}^{n} x_t \sin t\lambda\right|^2. \tag{3.76}$$

Effecting the substitution,

$$C_k = \frac{1}{n}\sum_{t=1}^{n-k} x_t x_{t+k}, \qquad k = 0, 1, \ldots, n-1, \tag{3.77}$$

we get the following expression for real sequences:

$$I_n(\lambda) = \frac{1}{2\pi}\left(C_0 + 2\sum_{k=1}^{n-1} C_k \cos k\lambda\right), \tag{3.78}$$

which is invariably used for computing numerically the values of the periodogram.

For the purposes of theoretical analyses the expression can be rewritten as

$$I_n(\lambda) = \frac{1}{2\pi}\sum_{k=-n+1}^{n-1} C_k\, \mathrm{e}^{-ik\lambda}, \tag{3.79}$$

where $C_k = C_{-k}$ is defined for $k < 0$, and where $(\mathrm{e}^{ik\lambda} + \mathrm{e}^{-ik\lambda})/2$ has been substituted for $\cos k\lambda$ in equation (3.78).

If we now compare formula (3.79) with the formula of spectral density, which is usually defined in the following form,

$$f(\lambda) = \frac{1}{2\pi} \sum_{t=-\infty}^{\infty} e^{-it\lambda} R(t), \tag{3.80}$$

it becomes clear that C_k can be regarded as a kind of estimator of covariance function $R(k)$, and that the periodogram can thus be viewed as an empirical estimator of spectral density.*)

Anděl [2], however, remarks that the periodogram need not be a generally consistent estimator of spectral density, and he claims that with density $f(\lambda)$, continuous, the periodogram can in limit cases (with $n \to \infty$) be regarded as its asymptotically unbiassed estimator.

Thus, if a large number of independent and sufficiently long realizations of random sequences are available, their periodograms and their arithmetic means are computed, which can approximately be regarded as estimates of spectral density $f(\lambda)$. But the greatest difficulty arises if just a single realization of random sequence is available. Since its periodogram need in no way be a sufficient estimator of spectral density, such numerical procedures must be sought that will yield better estimates.

The literature mentions a number of numerical methods of estimating spectral density based upon the theoretical fact that a certain transformation of the periodogram (viz. e. g. an integral of the product of a function and the periodogram) could produce both an asymptotically unbiassed estimator and, by contrast with a simple periodogram, a consistent estimator. This approach has resulted in estimators of spectral density of the following type:

$$f^*(\lambda) = C_0 w_0 + 2 \sum_{k=1}^{n-1} C_k w_k \cos k\lambda, \tag{3.81}$$

where C_k, $k = 0, \ldots, n-1$ are autocovariance coefficients, and coefficients $w_0, w_1, \ldots, w_{n-1}$, often referred to as weight coefficients, are selected with respect to certain algorithms. (The literature [2, 74] mentions, for example, the general Blackman–Tukey estimator, the Tukey–Hamming estimator, the Bartlett estimator, and the Parzen estimator).

*) For the spectral density of a stationary sequence to exist, it suffices for its covariance function that

$$\sum_{t=-\infty}^{\infty} |R(t)| < \infty .$$

The Parzen estimator appears to have proved the most appropriate. This estimator smooths the autocovariance function with weight coefficients w_k in the following form:

$$w_k = \frac{1}{2\pi}\left[1 - \frac{6k^2}{K^2}\left(1 - \frac{k}{K}\right)\right] \quad \text{for} \quad k = 0,1, \dots, \frac{K}{2},$$

$$w_k = \frac{1}{\pi}\left(1 - \frac{k}{K}\right)^3 \quad \text{for} \quad k = \frac{K}{2} + 1, \dots, K, \tag{3.82}$$

$$w_k = 0 \quad \text{for} \quad k > K,$$

where K is an even number invariably selected from within the range $n/6$ to $n/5$. The estimates of spectral density are recommended to be computed for frequencies

$$\lambda_j = \frac{\pi j}{K} \quad \text{for} \quad j = 0, 1, \dots, K. \tag{3.83}$$

It is an advantage of this estimator that the estimate of spectral density is then non-negative.

At present, such numerical procedures are being sought that could both yield satisfactory estimates of spectral density and also be effective from the point of view of the simplicity of computation. The requirement can thus be formulated as fast computation of the periodogram together with its simple smoothing with the help of weight coefficients.

The literature [40] also mentions other numerical smoothing methods, according to which not only autocorrelation functions, but also spectral densities, can be transformed with the help of weight functions. In this sense, weight functions are sometimes referred to as correlation, or spectral, windows.

The difficulties in computing spectral density from a limited number of observations of hydrological quantities arise basically from the fact that hydrological processes, apart from the regular (non-accidental, periodical) components, also exhibit accidental components, which are the result of the effect of fortuitous factors. The shares of these two types of components can in no way be estimated in advance. But there exist methods of statistical filtration, which provide adequate suplementary methodological means of analysing time series, particularly the means of ascertaining the periodic properties of the time series. Filtration is thus considered to be a particular case of a random variable estimator engaged in removing the accidental components from a given random sequence. The underlying concept here is that a given realization of a random sequence is a sum of both the random and the non-random variables in the form of an absolutely random sequence. The two types of components are separated with the help of special algorithms (filters), which can expose the composition of the original series as well as the probability properties of its components. Using a filter may, for instance, highlight the periodic components in a series.

The process of filtration can be elucidated with the help of a simple example of two random sequences, $X(t)$ and $Y(t)$, with realization at discrete time points

$$\left.\begin{array}{l} x(t-n), \ldots, x(t-1), x(t), x(t+1), \ldots, x(t+m), \\ y(t-n), \ldots, y(t-1), y(t), y(t+1), \ldots, y(t+m), \end{array}\right\} \tag{3.84}$$

where $m \geqq 0$. $X(t)$ will denote a random sequence of a useful signal; $Y(t)$ a random sequence of noise. Let us suppose that the two sequences cannot be examined separately, their realizations are thus unobtainable, so that only their sum is available, in the following form:

$$z(t-n), \ldots, z(t-1),$$

for which it holds that

$$z(t') = x(t') + y(t'), \qquad t - n \leqq t' \leqq t - 1. \tag{3.85}$$

Filtration involves finding the best estimator $X(t')$ of sequence $X(t)$ within the interval $t - n \leqq t' \leqq t + m$ on the basis of the knowledge of the past course of the sequence $z(t')$. With $m \geqq 0$, the filtration is linked with prediction (extrapolation); with $m < 0$, the filtration is retrospective.

From the problem of filtration presented above it follows that its essence consists in a function being found such that it is the best approximation to quantities $x(t+m)$, viz.

$$\hat{x}(t+m) = f[z(t-1), z(t-2), \ldots, z(t-n)]. \tag{3.86}$$

As far as the stationarity of the problem is concerned, it is assumed that the two random sequences, $x(t)$ and $y(t)$, are stationary, mutually uncorrelated, and that their expected values equal zero.

As in the case of prediction, the accuracy of filtration can be measured by the minimum of variance, viz.

$$\sigma^2_{m,n} = M\{x(t+m) - f[z(t-1), z(t-2), \ldots, z(t-n)]\}^2. \tag{3.87}$$

Finding a function (3.86) of a form for which (3.87) will be minimal, is a very complex task, which cannot be dealt with within the framework of the theory of correlation. As in the case of extrapolation, we therefore limit ourselves to linear approximation (linear filtration), and hence function (3.86) will assume the following form:

$$\hat{x}(t + m) = a_1 z(t - 1) + a_2 z(t - 2) + \dots + a_n z(t - n). \qquad (3.88)$$

The problem thus boils down to the task of finding such values of coefficients $a_1, a_2, \dots, a_n$ for which the variance (3.87), rewritten in the form

$$\sigma^2_{m,n} = M\left\{x(t + m) - \sum_{k=1}^{n} a_k z(t - k)\right\}^2, \qquad (3.89)$$

is minimal. This task is of course relatively simple: it can be shown that the mere knowledge of correlation functions $r_x(\tau)$, $r_y(\tau)$, and $r_z(\tau)$ will prove entirely sufficient.

The solution involves making use of a system of linear algebraic equations in the following form:

$$r_x(m + k) - \sum_{l=1}^{n} a_l r_z(k - l) = 0, \qquad k = 1, 2, \dots, n, \qquad (3.90)$$

which will provide us with the required coeffcients $a_1, a_2, \dots, a_n$.

The generation of the moving averages of a given sequence $z(t)$ may be viewed as a particular case of filtration. The generation of moving averages is practically a process of transition to a new sequence, $x(t)$, which is obtained from the original sequence if for example

$$x(t) = \sum_{k=-n}^{n} a_k z(t - k), \qquad (3.91)$$

where a_k denotes the weight coefficients selected according to a given rule.

Let us suppose that the sequence $z(t)$ is defined at all time points $t = \dots, -2, -1, 0, 1, 2, \dots$. According to expression (3.91), the new sequence, $x(t)$, is thus generated symmetrically with respect to every t from terms $z(t - n)$ to $z(t + n)$. Series $x(t)$ is often referred to as a filtered $z(t)$ series.

If $z(t)$ is a stationary random sequence, sequence $x(t)$ is also stationary. The generation of the moving averages will however change the correlation function of the two sequences of uncorrelated random variables into a correlated random sequence.

However, the examination of the effect of the moving averages (filters) can sometimes be very difficult, particularly if the probability properties of the original series $z(t)$ exhibit greater complexity. And this is also the reason why formulation of the relationship between the spectral densities of the two series, $x(t)$, $z(t)$, is sometimes interchanged when these problems are to be solved; for it can be shown that the spectral density of the filtered $x(t)$ series, which stresses the effect of the periodic component, can under certain conditions be achieved by the spectral density of the original $z(t)$ series being multiplied by the squared transfer function of filter a_k, i. e. that the following relationship holds:

$$s_x(\omega)|D(\omega)|^2 \, s_z(\omega) \,, \tag{3.92}$$

where the transfer function of the filter, $D(\omega)$, is defined as

$$D(\omega) = \sum_{k=-\infty}^{\infty} a_k \, \mathrm{e}^{-\mathrm{i}\omega k} \,. \tag{3.93}$$

In practice, we often come across filters of a truncated type, viz. $a_k = 0$ for $|k| > c$, where c stands for a finite number. If $a_k = a_{-k}$, the filter is referred to as symmetric; if $a_k = 0$ for $k < 0$, the filter is one-sided.

In our research we filtered hydrological and other geophysical series by generating moving averages with the help of weight coefficients in the form of binomial coefficients $\binom{k}{n}$, known from binomial distribution of probability (hence also the name "binomial filters").

We therefore first expressed the given terms of series Q_t in the following form:

$$Q_t = \bar{Q}_t + \varepsilon_t \,, \tag{3.94}$$

where $\bar{Q}_t$ represents the moving averages, and ε_t the random component (uncorrelated sequence with minimum dispersion).

The moving averages, $\bar{Q}_t$, were then generated according to the following formulae:

$$\left.\begin{aligned}
\bar{Q}_t^{(1)} &= \tfrac{1}{2}(Q_t + Q_{t+1}) && \text{1st degree of approximation,} \\
\bar{Q}_t^{(2)} &= \tfrac{1}{4}(Q_t + 2Q_{t+1} + Q_{t+2}) && \text{2nd degree of approximation,} \\
\bar{Q}_t^{(3)} &= \tfrac{1}{8}(Q_t + 3Q_{t+1} + 3Q_{t+2} + Q_{t+3}) && \text{3rd degree of approximation,} \\
&\vdots \\
\bar{Q}_t^{(k)} &= \frac{1}{2^k}\left[Q_t + kQ_{t+1} + \frac{k(k-1)}{2!} Q_{t+2} + \right. && k\text{-th degree of approximation.} \\
&\quad \left. + \frac{k(k-1)(k-2)}{3!} Q_{t+3} + \ldots \right]
\end{aligned}\right\} \tag{3.95}$$

Generating moving averages according to formulae (3.95) is not the only possible procedure. According to the character of the time series, other types of moving averages can also be constructed.

TABLE 1. Basic data of the set of long-term time series under examination

No. of series	Type of series	Place of observation	Country	River	Period of observation	Number of the elements of the series
1	flow	Norslund	Sweden	Dal	1853–1922	70
2	flow	Dnepropetrovsk	CIS	Dnepr	1882–1955	74
3	flow	Lotsmano-Kamenka	CIS	Dnepr	1818–1955	138
4	flow	Stein-Krems	Austria	Danube	1829–1960	132
5	flow	Orşova	Roumania	Danube	1838–1957	120
6	flow	Murchison	Australia	Goulburn	1882–1954	73
7	flow	Sjötörp-Vänersburg	Sweden	Göta	1808–1957	150
8	flow	Kamawha-Falls	USA (West Virginia)	Kanawha	1878–1957	80
9	flow	Kiewa	Australia	Kiewa	1886–1957	72
10	flow	Děčín	Czechoslovakia	Elbe	1851–1963	113
11	flow	Keokuk	USA (Iowa)	Mississipi	1879–1957	79
12	flow	St. Louis	USA (Missouri)	Mississipi	1861–1963	103
13	flow	Moravský Ján	Czechoslovakia	Morava	1895–1960	66
14	flow	Arad	Roumania	Mureş	1877–1955	79
15	flow	Albury	Australia (N.S. Wales)	Murray	1877–1950	74
16	flow	Smalininkai	USSR	Nemen	1812–1943	132
17	flow	Petrokrepost	USSR	Neva	1860–1935	76
18	flow	Greenville	Canada	Ottawa	1871–1959	89
19	flow	Basel	Switzerland	Rhine	1808–1951	144
20	flow	Ogdensburg	USA (N. York)	St. Lawrence	1861–1957	97
21	flow	Teddington	Great Britain	Thames	1884–1954	71
22	flow	Chattanooga	USA (Tennessee)	Tennessee	1875–1956	82
23	flow	Prague	Czechoslovakia	Vltava	1825–1966	142
24	precipitation	Děčín-Libverda	Czechoslovakia	1851	1851–1962	112
25	precipitation	Havlíčkův Brod	Czechoslovakia	–	1851–1962	112
26	cloudiness	Prague	Czechoslovakia	–	1861–1960	100
27	precipitation	Prague-Clementinum	Czechoslovakia	–	1851–1962	112
28	temperature	Prague-Clementinum	Czechoslovakia	–	1771–1965	195
29	sun spots	–	–	–	1749–1964	216

We applied the method of binomal filtering to a set of twenty-nine time series (quoted with their basic data in Table 1). Of these time series twenty-three were flow series and six various meteorological and other series (of precipitation, cloudiness, air temperature, sun spots), mostly of greater length.

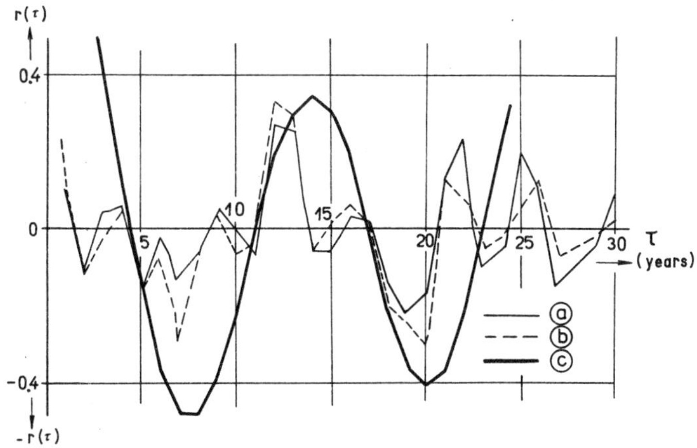

Fig. 6. Curves of correlation function of average annual flows in the Norslund profile on the river Dal (Sweden):
(a) autocorrelation function of the original 70-year series over the period of 1853–1922,
(b) average correlation function of the correlation functions of 50-year moving samples of the original series,
(c) correlation function of the binomially filtered original series (degree of filtration k = 20).

The set of the time series was assembled in order to include the largest possible number of the long series that were available. The set thus comprises series the length of which ranges between 66 years (the average annual flows in Moravský Ján in Moravia, Czechoslovakia) and 216 years (the average annual relative numbers of sun-spots); in all the cases the variables were average annual values.

Before filtration was carried out, the fundamental probability properties of all the time series had been analyzed; the moment characteristics of distributions had been computed as well as the sample autocorrelation functions and periodograms. We then constructed the filtered series, invariably in three variants of the degree of binomial filtration, viz. k = 10, 20, 30. For the series that had been filtered, the autocorrelation functions and the densities estimated had then to be computed again. The research also comprised a study of the properties of the series filtered related to a gradually raised degree of filtration, as well as a study of the problems linked with the stability of filtration.

Figure 6 shows an example of the computation of the correlation functions of average annual flows of the Swedish river Dal in the Norslund profile. A comparison is made of the curves of the autocorrelation function of the given series, the average correlation function derived from the set of sample autocorrelation functions, and the correlation function of the original series binomially

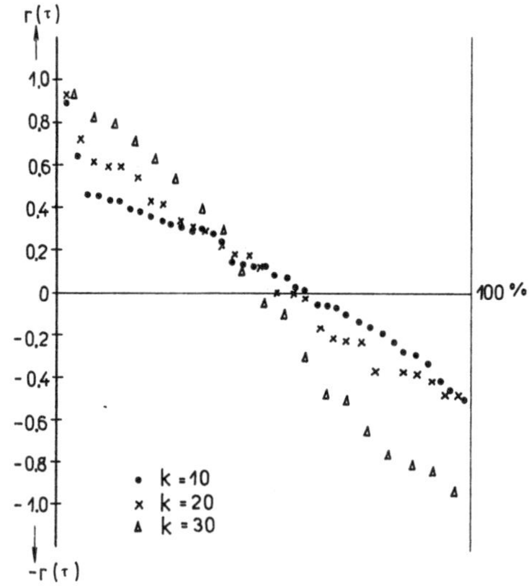

Fig. 7. Lines of transgression of the values of correlation functions of filtered flow series in the Norslund profile on the river Dal (Sweden) with various degree of filtration k.

filtered at the degree $k = 20$. Curves (a) and (b) do not differ substantially as far as the periodicity of variation, the occurrence of maxima and minima, and the amplitudes and their instantaneous values are concerned. Curve (c) is the most interesting; it is cleared of all short-term, mostly random, deviations and changes. It particularly highlights the existence of the periodic component of length about 12 to 14 years (with maximum and minima for τ equal to 14, and 8 and 20, respectively). The amplitudes of curve (c) are, in the given case, higher with all the degrees of filtration as compared with the autocorrelation function of the original series and the average correlation function. The comparative analysis thus points to an increase of autocorrelations with both the generation of the moving averages and the smoothing of the series with the help of a binomial filter.

The growth of autocorrelations is even more marked in Fig. 7, which shows the curves of the transgression of the values of the correlation functions of the series filtered at different degrees of filtration.

In examining periodicity we concentrated, apart from correlation functions, on the estimators of spectral densities of the series filtered, which provide a better possibility of detecting the existing periodic components. It was however found that the spectra of the filtered series of a larger set need not have a simple and always similar curve either. This can be explained by the specific probability and genetic properties of the individual series. Moreover, the effect of the degree of filtration related to various lengths of a given historical series can also manifest itself to a certain extent. In our research we were fully aware of this effect, which however proved to be rather difficult to estimate qualitatively.

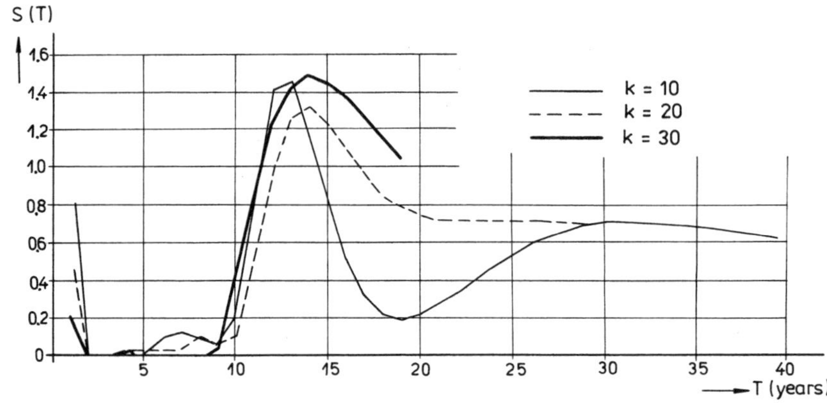

Fig. 8. Spectral density function for various degrees of filtration (Dal–Norslund).

The individual spectra of the filtered series usually exhibit ragged curves in the region of very short periods. The following part of the spectrum then often has a more pronounced narrow-zone character, which enables us to infer the exist-

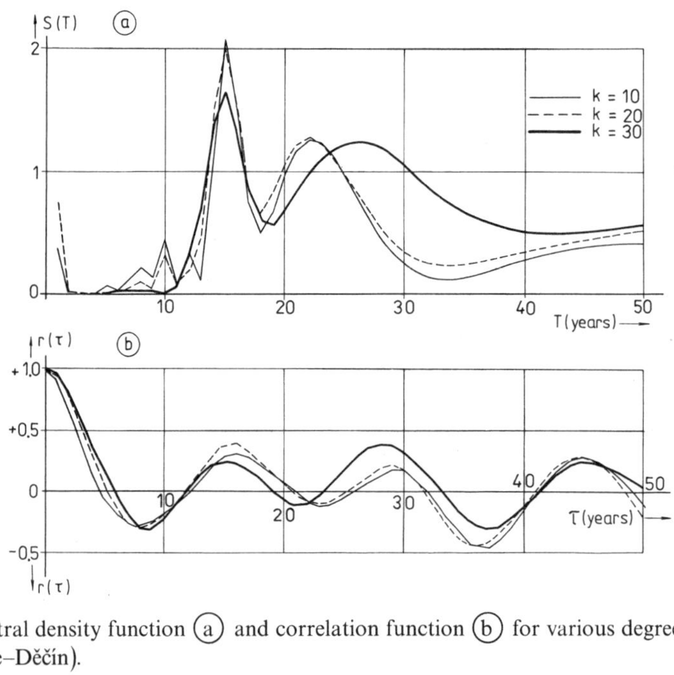

Fig. 10. Spectral density function (a) and correlation function (b) for various degrees of filtration (Elbe–Děčín).

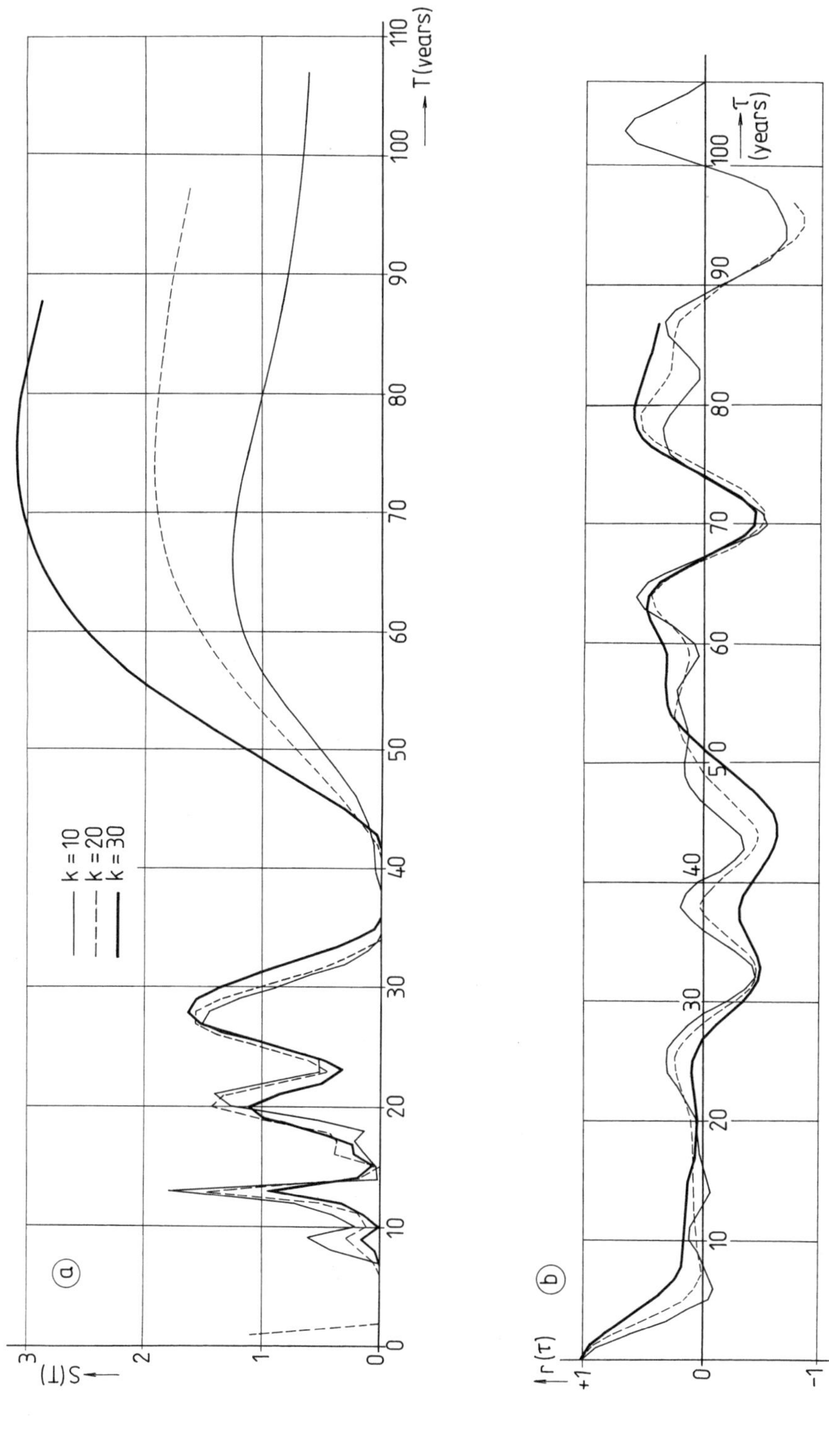

Fig. 9. Spectral density function (a) and correlation function (b) for various degrees of filtration (Dněpr – Lotsmanska Kamenka).

ence of a medium-long period in the series. This part of the spectrum can also be composed of several sections, which confirms the information acquired by means of other methodological procedures, namely, that the series of hydrological variables can include several periods.

TABLE 2. Survey of the more significant periodic components of the curves of the functions of spectral density

Series No.	Region of periods					
	short and medium-long			long		
	degree of filtration			degree of filtration		
	10	20	30	10	20	30
1	13	14	14	–	–	–
2	9, 12, 13	21	21	–	–	–
3	9, 13, 21	9, 13, 20	9, 13, 20	27, 66	27, 75	28, 75
4	9, 12, 16, 24	9, 12, 16, 24	12, 16, 24	50	52	52
5	9, 15, 21	9, 15, 21	15, 21	32	32	33
6	14	13	13	39	–	–
7	7, 11, 15, 21	11, 16, 20	12, 15, 20	31, 45	30, 46	30, 47
8	9, 16	9, 16	9, 18	27	–	–
9	(8), 12, (20)	(12)	11	–	–	–
10	10, 15, 22	10, 15, 22	15	–	–	26
11	7, 12, 19	(9), (13), 18	(7), 14	38	–	–
12	7, 15, 23	(8), 12, 15, 22	(11), 21	39	36	–
13	(7), 25	–	–	–	–	–
14	9, 14	10, 14	(11)	29	33	–
15	(10), 14, 17	(9), 13	(7), (12)	31	32	–
16	8, 12, 15, (20)	9, 12, 16, 19	(10), 12, 16, 18	26	26	26
17	12	12	11	28	28	–
18	(9), 13, 19	(10), (12), 18	(16)	32	33	31
19	9, 12, (15), 19, (22)	(7, 9), 12, 15, 18, 22	(10), 12, 18, 22	30	29, 84	30
20	10, 15, 20	14, 20	13	35	35	35
21	(8), 12	(7), 12	–	–	–	–
22	9, 15	15	13	39	35	–
23	10, (12), 15, 23	(10), 15, 23	10, 15, 23	44	45	50
24	10, (13), 16,	10, (12), 16	(10), (12), (17)	(26)	29	29
25	10, 16, 18	(12), 16, (25)	(11), 17, (23)	26, 47	47	46
26	20	19	15	–	–	–
27	8, (10, 12, 14), (21)	(8), 14, (20)	13, (18)	48	50	37
28	(9), 14, 18	14, 18	14, 18	38	38	38
29	10, 21	(12, 16, 20)	11, (20)	28, 41, (56), 90	27, 40, (54), 88	26,86

Figure 8 is an example of a simple curve of spectral density of the filtered Dal-Norslund series at the three degrees of filtration mentioned above. The maximum ordinates occur in the region of $T = 13-14$ years. These curves show that in a number of cases a lower degree of filtration will suffice to demonstrate the periodic component. Higher degrees of filtration thus need not invariably lead to new information.

Figure 9 shows a much more complex curve of spectral density of the filtered series of average annual flows of the Dnepr in the L. Kamenka Commenwelth of Independent States (CIS) profile, again at three degrees of filtration. Three more significant periods can be identified in these curves, viz. $T = 13$, 20–22, and 27–28 years. The broad-zone character of the following section of the curve is also interesting.

Figure 10 shows the curve of spectral density of the filtered series of average annual flows of the Elbe in the Děčín (Czechoslovakia) profile, where two extremes in the region of medium-long periods can be identified. The relatively most pronounced extreme corresponds to $T = 15$ years.

Table 2 presents a survey of the more pronounced periodic components in the set of 29 selected time series. In all the cases, spectral densities were computed for the binomially filtered series. The periods are divided into two groups:

a) short and medium-long periods (up to 25 years, incl.);

b) long periods (26 years and more).

Apart from the periods corresponding to the more conspicuous values of the spectrum, Table 2 also presents other periods, which correspond to the less pronounced ordinates of spectral density (in brackets). In the columns, the values are further differentiated according to the degree of filtration of the series.

From the lengths of the periods in the whole set of twenty-nine time series a histogram was plotted (Fig. 11) for the three-year classes $III_{f(T)}$ of the lengths of the periods. The periods of 10–12 years, and 13–15 years were found to be relatively the most frequent in the given set.

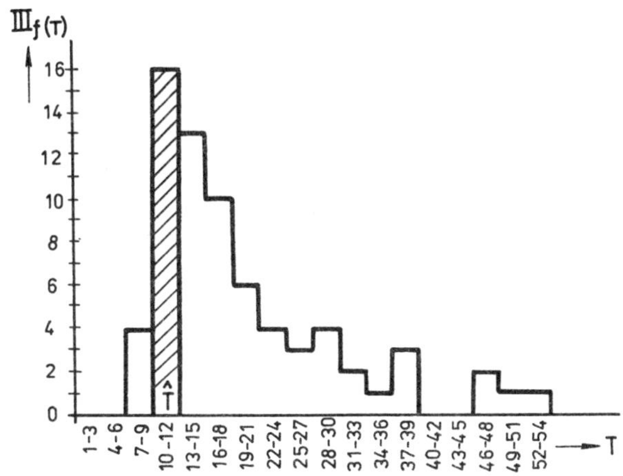

Fig. 11. Histogram of class frequencies of the occurrence of various periods in a set of twenty nine filtered series.

As far as the overall assessment of the methods of filtration is concerned, it can be claimed that these methods are adequate and effective instruments for the analysis of the periodical properties of time series. The research carried out has however also revealed some problems of statistical analysis, which will require some attention. Basically, these problems follow from the complex probability (particularly autocorrelative) properties of some of the time series. As far as the methods of filtration are concerned, the greatest problems were posed by the estimation of the weight coefficients and the degree of filtration. And besides, it is obvious that with the length of the series avilable being limited, no high degree of filtration can be chosen, because the series filtered gets shorter and its analysis is thus made more difficult.

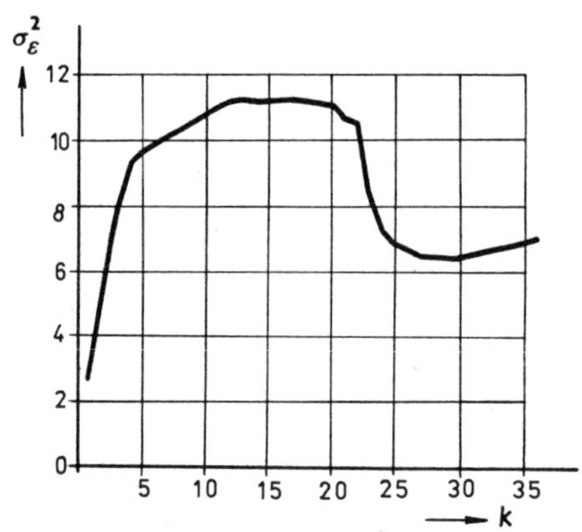

Fig. 12. Dependence of residual variance on the degree of filtration of an annual flow series of the river Elbe at Děčín.

In some cases problems of examining the dependence of residual variance upon the degree of filtration may prove to be rather complex. Consider the example presented in Fig. 12 showing the dependence of residual variance σ_ε^2 upon the gradually raised degree of binomal filtration $(k = 1, 2, \ldots, 36)$ for the flow series of the Elbe in Děčín (Czechoslovakia). Minimum variance occurred as early as with $k = 1$, it then gradually rose until the maximum values were reached in a relatively wide region, viz. $k = 10$–20. The initial shape of the dependence curve can be accounted for by the fact that, with the series gradually smoothed, its dispersion will grow less and residual variance will increase. But in the broad region of the extremes of σ_ε^2, the effect of filtration is indistinct. And the relatively marked decline of σ_ε^2 with $k > 20$ is also hard to explain.

Difficulties were encountered in the assessment of the autocorrelative properties of residual deviations. In some cases the curves of the autocorrelation functions of these deviations manifested significant values, which moreover could not be compared due to the different degrees of filtration.

The problems indicated above will require further research.

The applications of spectral analysis to hydrological series have been treated by a number of authors (e. g. Yevjevich [121], Buchtele [21] in Czechoslovakia); the correlation functions and the corresponding spectral densities of the annual flow series were analyzed by Nacházel and Patera [85]; the mutual relationship between the periodograms and the spectral densities of the annual flow series estimated were described in detail by Anděl [2], Anděl and Balek [4, 5] and others.

All the works quoted above prove that at present, spectral analysis is quite an elaborate methodological instrument for the assessment of the periodic properties of time series. Viewed from this point, spectral analysis is an indispensable initial step towards the construction of the periodic models of time series. The numerical applications, however, also show that the estimation of spectral densities from periodograms requires particular experience and skills to facilitate the computations. The Parzen formula has in most cases proved itself in this respect.

3.4 Computation of point and interval estimates of parameters

In Chapter 1 we stated that the computation of point and interval estimates of parameters on the basis of the knowledge of the properties of the samples is one of the fundamental methodological procedures of the process of estimation. Although the initial requirements for their application may be similar, point estimation and interval estimation differ quite considerably. Whereas point estimation is a process of estimating a population parameter with a single number, an interval estimate is a range of values used to estimate a parameter; the parameter is thus estimated to be within a range of values.

The application of the two methodological procedures, point and interval estimation, depends primarily upon the nature of the problem to be tackled. The methods of point estimation (e. g. the well-known moments method and the method of maximum likelihood) are receiving considerable theoretical attention. These methods have found wide practical application where the solution of a problem is to be built upon a single estimated design value of parameters (e. g. in the designs of storage reservoirs, which are invariably based upon a single design value of the parameters of a flow series).

A disadvantage of point estimation is that it does not allow assessment of the precision of the estimate. This drawback is removed by interval estimation, which can provide an answer to the question concerning the admissible estimating error. Interval estimation has enjoyed a revival only recently, thanks to the development of the mathematical modelling of random sequences, the output parameters of which are verified with the help of confidence intervals.

We will now discuss the essence of the two methods.

For the mean of a given population with Gaussian distribution equation (3.47) holds, according to which the mean of sample means equals the population mean. If however only a single sample mean is known, it stands to reason that we will risk the least error, as far as the population mean is concerned, if the given sample mean is assumed to be equal to the mean of all the sample means. From this consideration it immediately follows that it is the sample mean, $\bar{x}$, that is the best point estimate of an unknown mean of a population mean μ. This estimate can thus be written in the following form:

$$\dot{\mu} = \bar{x} \,. \tag{3.96}$$

A similar consideration will bring us to the point estimate of dispersion or standard deviation of a population. From equation (3.18) the following relationship follows for the unknown variance σ^2:

$$\sigma^2 = \frac{n}{n-1} E(s^2) \,. \tag{3.97}$$

If the variance of a single given sample, s^2, is known, it is again logical that the least error will be made, as far as the estimate of σ^2 is concerned, if the given variance is considered to be the mean of all the sample variances. Thus, if s^2 is substituted for $E(s^2)$ in equation (3.97), the point estimate of σ^2 will have the following form:

$$\dot{\sigma}^2 = \frac{n}{n-1} s^2 = \frac{n}{n-1} \frac{1}{n} \sum_{i=1}^{n} (x_i - \bar{x})^2 = \frac{1}{n-1} \sum_{i=1}^{n} (x_i - \bar{x})^2 = S^2 \,. \tag{3.98}$$

We thus arrive at the expression of sample variance S^2, which is at the same time an unbiassed estimator of σ^2, as already mentioned above.

For the point estimation of the standard deviation of a population , equation (3.98) yields the following relationship:

$$\dot{\sigma} = \sqrt{\frac{1}{n-1} \sum_{i=1}^{n} (x_i - \bar{x})^2} \,. \tag{3.99}$$

Point estimation can become a rather difficult task, particularly if parameters are to be computed of random sequences that do not exhibit Gaussian distribution. And it is sequences of this type that are most frequently dealt with in hydrology and in water engineering. In this case the problems are mainly due to the fact that the estimators are invariably biassed; due account should therefore also be taken of the non-negligible systematic error. Another difficulty can consist in the numerical exaction involved in the best estimator, or in the fact that even the best estimator can be biassed, and that it is only in the limit case (viz. $n \to \infty$) that it will approximate to an unbiassed estimator. In such cases the problems are avoided by resorting to another methodological procedure and assessing its dependability.

These difficulties are dealt with in the following chapters of this book.

For the computation of interval estimates a knowledge of the probability distribution of the respective sample characteristic is indispensable. The essence of this estimation consists in a specific interval of the distribution of the characteristic being selected, wide enough to contain the unknown parameter.

Let us again assume a population with Gausssian distribution. The distribution of the sample means is then also normal, and the two-sided interval including radom variables $\bar{x}$ with probability 1–2p can easily be derived in the following form:

$$E(\bar{x}) - t'_p\,\sigma(\bar{x}) < \bar{x} < E(\bar{x}) + t'_p\,\sigma(\bar{x}) \tag{3.100}$$

where t'_p stands for the value (quantile) of standardized normal quantity t', which is given by equation (3.50).

In view of expressions (3.47) and (3.49), inequality (3.100) can be rewritten in the following form:

$$\mu - t'_p \frac{\sigma}{\sqrt{n}} < \bar{x} < \mu + t'_p \frac{\sigma}{\sqrt{n}}, \tag{3.101}$$

from which an explicit expression for the mean of the population, μ, can easily be obtained, viz.

$$\bar{x} - t'_p \frac{\sigma}{\sqrt{n}} < \mu < \bar{x} + t'_p \frac{\sigma}{\sqrt{n}}. \tag{3.102}$$

The standard deviation, σ, of the population is actually unknown. The point estimate $s\sqrt{n/(n-1)}$ will therefore substitute for it, and inequality (3.102) can be rewritten in the following form:

$$\bar{x} - t_p \frac{s}{\sqrt{n-1}} < \mu < \bar{x} + t_p \frac{s}{\sqrt{n-1}}, \tag{3.103}$$

where t_p denotes the value (quantile) of quantity t exhibiting distribution t with $n - 1$ degrees of freedom, and given by equation (3.51).

In practice, use is also made, apart from the two-sided confidence interval, of single-sided confidence intervals, either top- or bottom-limited. For the top-limited confidence interval, and with the variance of the population unknown, the following inequality holds:

$$\mu < \bar{x} + t_p \frac{s}{\sqrt{n-1}} ; \tag{3.104}$$

and for the bottom-limited confidence interval it holds that

$$\mu > \bar{x} - t_p \frac{s}{\sqrt{n-1}} . \tag{3.105}$$

Analogous procedures can be used to derive interval estimates of the variance of a population with Gaussian distribution in the following form:

$$\frac{n}{\chi^2_{p_2}} s^2 < \sigma^2 < \frac{n}{\chi^2_{p_1}} s^2 , \tag{3.106}$$

where $\chi^2_{p_1}$ and $\chi^2_{p_2}$ represent the quantiles of distribution χ^2 with $n - 1$ degrees of freedom for probabilities p_1 and p_2 (see Fig. 13).

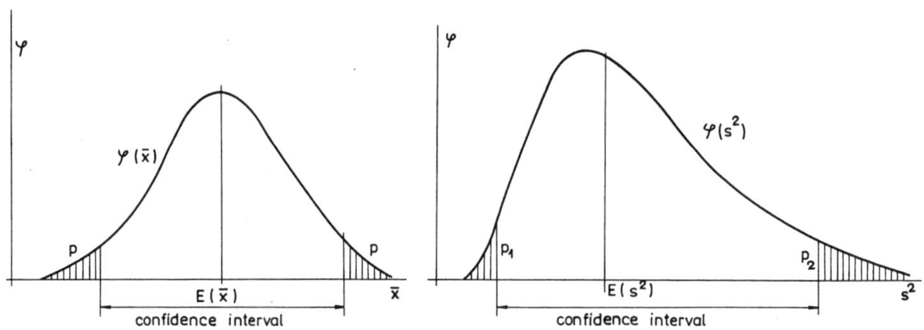

Fig. 13. Two-sided confidence interval for mean μ and variance σ^2.

For a top-limited single-sided confidence interval the following inequality holds:

$$\sigma^2 < \frac{n}{\chi^2_{p_1}} s^2 . \tag{3.107}$$

A bottom-limited confidence interval is of no practical use.

Extracting the square roots of inequalities (3.106) and (3.107) will furnish us with confidence intervals for the respective standard deviations.

The literature [35, 114] also quotes confidence intervals for the estimation of other parameters.

For water-engineering computations, the most important parameter is the estimation of the correlation coefficient of a population with the help of the sample coefficient of correlation, r. This is a random variable lacking Gaussian distribution. Use is therefore made of the computation of the confidence interval of the transformed random variable devised by R. Fisher, viz.

$$z = \frac{1}{2} \ln \frac{1 + r}{1 - r}, \text{*)} \tag{3.108}$$

which has, for a proportionally great n (approximately, $n \geqq 10$, unless $|\varrho|$ approximates to unity) an approximately normal distribution with the expected value

$$E(z) = \frac{1}{2} \ln \frac{1 + \varrho}{1 - \varrho}, \tag{3.109}$$

and standard deviation

$$\sigma(z) = \frac{1}{\sqrt{n - 3}}, \tag{3.110}$$

where n is the size of the random sample from which the correlation coefficient has been computed.

The two-sided confidence interval of quantity z can, in accordance with equations (3.100) and (3.102), be constructed in the following form:

$$z - t'_p \frac{1}{\sqrt{n - 3}} < E(z) < z + t'_p \frac{1}{\sqrt{n - 3}}, \tag{3.111}$$

where t'_p again denotes the quantile of the standardized normal random variable for probability p. Let us note that the confidence interval covers the unknown value of $E(z)$ with probability 1–2p.

The interval estimation of the unknown value of ϱ thus involves first computing the transformed value, z, from the known value of r, and then the upper and the lower limits of interval (3.111), the values of which are then converted back to the limits of the estimated parameter ϱ. To facilitate computation, the literature provides auxiliary values for the determination of $z = f(r)$, or $E(z) = f(\varrho)$.

The construction of interval estimates is linked with a number of pitfalls, due to the fact that for some of the more complex types of probability distribution the required analytical relationships between the distribution of sample charac-

---

*) ln denotes the natural logarithm.

teristics and parameters have so far not been derived. As with point estimation, problems arise with the biassed estimators. Despite these drawbacks, we can however claim that interval estimates, provided of course they can be constructed at all, are a valuable methodological instrument wherever an assessment of the accuracy of the estimate is to be undertaken using the length of the interval.

4
Estimation of parameters by the moments method

4.1 Principles of the moments method and the application of simulation models of random sequences to estimation

The principle of the moments method should be evident from what has been said in the preceding chapter, in which the problems of point estimation of the unknown parameters were discussed. It can be claimed that owing to its computational simplicity, elaborateness and universality, the moments method has become the most frequently used method of point estimation based upon mutual comparison between the moments of the population and the moments of the sample. The comparison aims at finding the relationships generally obtaining between the characteristics and the parameters for the given probability properties of a population. And it is upon these relationships that the success of parameter estimation is dependent.

Let us now return to Fig. 2 (p. 27) presenting a general case of the relationship between the biassed curve of the expected values of characteristics $E(u)$ and parameter u_0. If an analytic expression can be found for this relationship, then the unknown parameter u_0 is estimated as the sum of $E(u)$ and the corresponding systematic error Δ (according to equation (3.11)), viz.

$$u_0 = E(u) + \Delta \,. \tag{4.1}$$

In cases where (4.1) cannot be expressed analytically, use can be made of the simulation models of random sequences generated for the required probability properties of the population. This methodological procedure is justified primarily by fact that the technique of modelling random sequences on computers has already been satisfactorily elaborated, so that its application does not in practice pose any problems of a methodological character.

With this approach to parameter estimation, random sequences (of finite length) simulate the population (theoretically of infinite length) to a certain degree of reliability. It is therefore the basic requirement for the generated random sequence that it should be long enough for the random deviations of its output parameters to be minimal.*)

Generating random samples from the sequence modelled is the second step in the process of estimation. Random samples can be generated in several ways, the

TABLE 3.
Moments method
Pearson's IIIrd type distribution

Inputs: $\bar{x} = 1.00$, $C_v = 0.75$, $C_s = 1.50$		Outputs: $\bar{x} = 0.995\,8$, $C_v = 0.774\,6$, $C_s = 1.661\,2$	
Characteristics of 500 samples			
	$n = 20$	$n = 30$	$n = 40$
$E(\bar{x})$	0.997 8	0.998 5	0.994 9
$\sigma(\bar{x})$	0.174 4	0.141 6	0.121 9
$C_v(\bar{x})$	0.174 8	0.141 8	0.122 5
$C_s(\bar{x})$	0.505 3	0.132 2	0.244 3
$\bar{x}_{max}$	1.642 7	1.433 2	1.417 8
$\bar{x}_{97.5}$	1.380 8	1.300 0	1.272 0
$\bar{x}_{2.5}$	0.692 3	0.703 3	0.750 3
$\bar{x}_{min}$	0.460 0	0.598 8	0.685 0
$E(C_v)$	0.743 2	0.750 9	0.754 9
$\sigma(C_v)$	0.137 3	0.119 7	0.102 5
$C_v(C_v)$	0.184 8	0.159 4	0.135 8
$C_s(C_v)$	0.480 9	0.491 4	0.354 6
$C_{v,max}$	1.242 5	1.238 7	1.065 5
$C_{v,97.5}$	1.034 3	1.009 4	0.985 5
$C_{v,2.5}$	0.514 4	0.548 1	0.586 6
$C_{v,min}$	0.439 6	0.439 9	0.514 1
$E(C_s)$	1.053 1	1.197 3	1.286 2
$\sigma(C_s)$	0.586 9	0.582 4	0.551 1
$C_v(C_s)$	0.557 3	0.486 4	0.428 4
$C_s(C_s)$	0.679 1	0.713 0	0.771 7
$C_{s,max}$	3.086 6	3.210 7	3.386 0
$C_{s,97.5}$	2.512 6	2.617 2	2.519 9
$C_{s,2.5}$	0.082 3	0.267 4	0.485 4
$C_{s,min}$	−0.253 2	0.018 4	0.271 4

*) The results of our research show (see Section 4.3 and Chapters 8 to 10) that random sequences should be modelled in the lengths of at least 1 000 to 10 000 terms, in accordance with the values of their input parameters and the character of the sequences (viz. average annual flow sequences, maximum flood flows etc.). Rozhdestvenskiĭ [98] models 50 000-year series in order to obtain reliable estimates of systematic errors.

simplest being to select randomly the starting point of each sample [78]. And samples can vary in length, according to actual needs. What however matters is that the number of samples should be sufficiently large, in order that their behaviour and their probability properties may reliably be detected.

TABLE 4.

Moments method
Pearson's IIIrd type distribution

Inputs:		Outputs:	
$\bar{x} = 1.00$		$\bar{x} = 1.021\,4$	
$C_v = 1.00$		$C_v = 1.029\,3$	
$C_s = 2.00$		$C_s = 2.084\,2$	
Characteristics of 500 samples			
	$n = 20$	$n = 30$	$n = 40$
$E(\bar{x})$	1.023 2	1.033 0	1.028 3
$\sigma(\bar{x})$	0.227 1	0.180 5	0.156 5
$C_v(\bar{x})$	0.222 0	0.174 7	0.152 2
$C_s(\bar{x})$	0.392 1	0.186 4	0.219 7
$\bar{x}_{max}$	1.706 3	1.664 0	1.518 9
$\bar{x}_{97.5}$	1.493 2	1.415 7	1.360 6
$\bar{x}_{2.5}$	0.647 6	0.712 5	0.756 4
$\bar{x}_{min}$	0.465 7	0.534 7	0.574 0
$E(C_v)$	0.981 3	0.991 9	1.002 5
$\sigma(C_v)$	0.191 2	0.166 5	0.147 0
$C_v(C_v)$	0.194 9	0.167 8	0.146 6
$C_s(C_v)$	0.727 0	0.616 8	0.439 3
$C_{v,max}$	2.018 2	1.574 0	1.505 0
$C_{v,97.5}$	1.404 3	1.376 6	1.325 0
$C_{v,2.5}$	0.671 3	0.735 3	0.762 7
$C_{v,min}$	0.541 4	0.566 6	0.640 6
$E(C_s)$	1.315 8	1.478 2	1.611 0
$\sigma(C_s)$	0.607 7	0.602 7	0.614 6
$C_v(C_s)$	0.461 9	0.407 7	0.381 5
$C_s(C_s)$	0.485 5	0.585 4	0.605 5
$C_{s,max}$	3.338 9	3.243 9	3.736 1
$C_{s,97.5}$	2.690 9	2.888 3	2.963 7
$C_{s,2.5}$	0.318 7	0.518 2	0.667 8
$C_{s,min}$	−0.328 1	0.354 3	0.443 8

Computation of the characteristics of all the samples is the third step. If a computer programme is to be constructed, it is advisable that the programme should include the computation of all the basic characteristics: sample means, sample standard deviations and coefficients of variance, sample coefficients of

TABLE 5.

Moments method
Pearson's IIIrd type distribution

Inputs: $\bar{x} = 1.00$, $C_v = 1.20$, $C_s = 2.40$		Outputs: $\bar{x} = 1.022\,6$, $C_v = 1.219\,9$, $C_s = 2.477\,2$	
Characteristics of 500 samples			
	$n = 20$	$n = 30$	$n = 40$
$E(\bar{x})$	1.023 8	1.018 8	1.022 1
$\sigma(\bar{x})$	0.282 3	0.225 0	0.192 6
$C_v(\bar{x})$	0.275 8	0.220 9	0.188 4
$C_s(\bar{x})$	0.438 5	0.390 5	0.262 7
$\bar{x}_{max}$	1.913 5	1.822 8	1.651 8
$\bar{x}_{97.5}$	1.670 4	1.509 9	1.397 8
$\bar{x}_{2.5}$	0.564 3	0.636 5	0.688 3
$\bar{x}_{min}$	0.310 3	0.459 6	0.558 3
$E(C_v)$	1.150 0	1.168 2	1.180 0
$\sigma(C_v)$	0.238 3	0.189 2	0.180 5
$C_v(C_v)$	0.207 2	0.162 0	0.153 0
$C_s(C_v)$	0.859 5	0.551 5	0.594 9
$C_{v,max}$	2.167 6	1.845 4	1.771 3
$C_{v,97.5}$	1.719 7	1.595 7	1.595 7
$C_{v,2.5}$	0.762 5	0.860 8	0.878 9
$C_{v,min}$	0.635 0	0.711 4	0.710 1
$E(C_s)$	1.548 1	1.752 0	1.869 5
$\sigma(C_s)$	0.664 4	0.659 5	0.674 6
$C_v(C_s)$	0.429 2	0.376 4	0.360 8
$C_s(C_s)$	0.589 6	0.729 0	0.716 6
$C_{s,max}$	3.584 3	4.169 6	4.243 2
$C_{s,97.5}$	3.141 8	3.376 9	3.473 1
$C_{s,2.5}$	0.492 5	0.705 2	0.860 1
$C_{s,min}$	0.060 6	0.499 3	0.351 5

asymmetry, and also sample coefficients of autocorrelation. The moment characteristics must then be computed for the sets of these characteristics (always for the given n), for example, for the set of the sample means the expected value of these means, the standard deviation, the coefficient of variation, the coeffcient of asymmetry etc.

Tables 3, 4 and 5 give examples of statistical processing of a set of 500 characteristics derived from 10 000-term random sequences with the Pearson distribution of the IIIrd type.

Statistical processing provides us with several results: the systematic error (related to the length of the sample) can easily be ascertained for each parameter; if necessary, the extreme values of the set of characteristics can be computed; confidence intervals can be derived, and the probability distribution of the characteristics graphically represented (Fig. 14a).

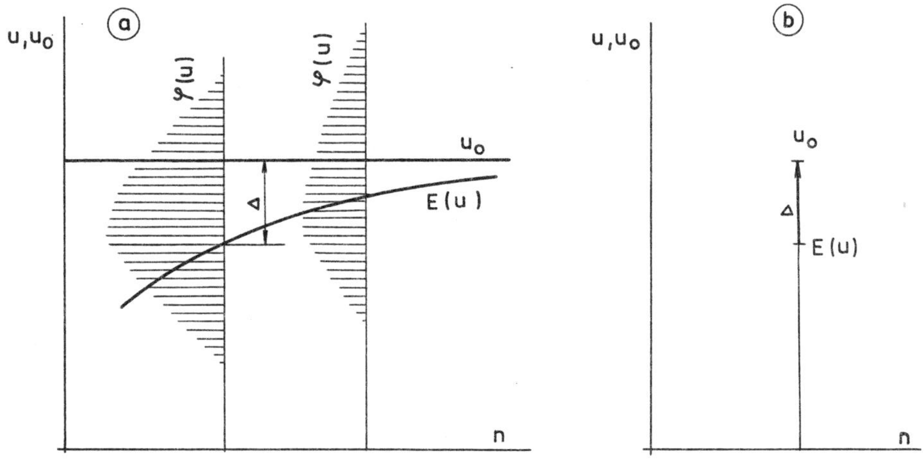

Fig. 14. Estimation of the parameters of a universe:
(a) model solution making use of random sequence,
(b) estimation on the basis of a single sample.

In practice, estimation proves to be difficult due to the fact that a single sample is invariably given and the precise probability distribution of the population is unknown. Unless the random errors of the individual characteristics can be assessed, then, in accordance with our considerations concerning point estimation, we proceed so that the characteristic computed is regarded as the expected value of the set of all the characteristics, and the systematic error (Fig. 14b) found with the help of the method of modelling is added to it.

With this procedure, the type of the probability distribution of the population is assumed in accordance with the experience gained from the analyses of longer real sequences. The systematic error must correspond to the length of the given sample.

From the procedure described above it can be seen that the possibility of using simulation models of random sequences to estimate the parameters of various types of series using the moments method will also expose some of the drawbacks involved. The computation of systematic errors from the set of random samples is numerically relatively exacting and it necessitates using efficient computer technology unless diagrams enabling quick determination of these errors are available. The bias of the estimators itself is often considered to be one of the disadvantages of the method; but biassed estimators can unfortunately also be found in methods providing efficient estimates. This is why attention should be given to systematic errors and, if necessary, the biassed expected values of the characteristics corrected to unbiassed parameters.

In the critical appraisal of the moments method, the problem of the property of the variance of the parameters estimated, corrected by the systematic error added, must not be omitted. Since this variance can increase, the question arises whether the deterioration of this property is offset by the bias of the estimator decreasing. It thus follows that none of the properties of the estimator should a priori be given preference, but that the given type of estimator should be carefully considered with due account taken of all its properties, and decisions should not be made until that time.

4.2 Estimation of parameters of populations with various probability distributions

Owing to the great advantages of simulation models of random sequences, the problems of the theory of estimation have recently started to be paid attention in the ex-Soviet Union, the USA, Canada and elsewhere. Attention has been focused on both the purely theoretical aspects and on the concrete methods of estimating sequence with various distributions of probability, the occurence of which is particularly frequent in the field of technology. Researchers aimed their first efforts at deriving the respective diagrams, which greatly facilitate estimation as compared with the modelling methods of solution. Progress continues to be made: the methods making automated processing of estimates possible have been rapidly developing and adequate analytic expressions are being sought for the systematic errors found in random sequence, because these expressions are so much more computer-friendly than diagrams.

In the ex-Soviet Union, Rozhdestvenskiĭ [98] made a valuable contribution in modelling sets of 50 000-terms series exhibiting the Pearson distribution of the IIIrd type and the three-parameter gamma distribution (in the CIS often referred to as Kritskiĭ–Menkeľ distribution) for various combinations of input parameters. From these series Rozhdestvenskiĭ generated up to 200 years long random samples, examined the bias of the characteristics, and constructed diagrams for rapid estimation of the unbiassed parameters.

Rozhdestvenskiĭ's diagrams are shown in Figs 15 and 16 (see App.). They are particularly easy to apply: for the given values of the sample coefficients of variation C_v, coefficient of asymmetry C_s, ratio C_s/C_v, coefficient of correlation between the neighbouring terms of the series, and the length of the sample, n, the user can readily read the unbiassed estimate of the respective parameter. As we have already mentioned in Section 4.1, the correction however involves the problem of the variance of the parameters estimated.

In keeping with the needs of practice, our research focused on the estimation of the parameters of the series with triparametric log-normal distribution and logarithmic Pearson's distribution of the IIIrd type [75].

4.2.1 *Estimation of parameters of a population with log-normal distribution*

In designing the model of random sequences with triparametric log-normal distribution characterized by three parameters (viz. the expected value $\bar{y}$, variance σ_y^2, and the minimum term x_0), we concentrated upon the basic task, i. e. to compute the input statistical parameters of the logarithms of variables

$$y = \ln (x - x_0) \tag{4.2}$$

for variables x with the given parameters.

For the general triparametric log-normal distribution the relationships are quoted for example in the Czechoslovak statistical literature [114, 117]; in other literature, this type has recently been dealt with in two interesting periodical articles [22, 38].

The mutual functional relationships between the parameters of variables x and the parameters of variables y are given by the following expressions [22]:

$$\bar{x} = x_0 + \exp (\bar{y} + \sigma_y^2/2)\,, \tag{4.3}$$

$$\sigma_x^2 = [\exp (\sigma_y^2) - 1] \exp (2\bar{y} + \sigma_y^2)\,, \tag{4.4}$$

$$C_{s,x} = \frac{\exp (3\sigma_y^2) - 3 \exp (\sigma_y^2 + 2)}{[\exp (\sigma_y^2) - 1]^{3/2}}\,. \tag{4.5}$$

Equation (4.5) gives only a single real root σ_y for the given positive coefficient of asymmetry $C_{s,x}$. The root is computed from the equation

$$\Phi = [\beta + (\beta^2 - 1)^{1/2}]^{1/3} + [\beta - (\beta^2 - 1)^{1/2}]^{1/3} - 1\,, \tag{4.6}$$

where

$$\beta = 1 + C_{s,x}^2/2\,, \tag{4.7}$$

$$\Phi = \exp(\sigma_y^2)\,. \tag{4.8}$$

Substituting (4.8) into (4.5) we get

$$C_{s,x} = (\Phi - 1)^{1/2}(\Phi + 2)\,, \tag{4.9}$$

from which we can easily express Φ for the given $C_{s,x}$, then σ_y from (4.8), and $\bar{y}$ from (4.4).

TABLE 6. Input parameters of models with triparametric log-normal distribution

Parameters of variables x				Parameters of logarithms of variables x		
$\bar{x}$	$C_{v,x}$	$C_{s,x}$	x_0	$\bar{y}$	σ_y^2	σ_y
1	0.50	$C_{v,x} = 0.50$	−2.013	1.089	0.026 8	0.164
		$2C_{v,x} = 1.00$	−0.558	0.395	0.098 2	0.314
		$3C_{v,x} = 1.50$	−0.072	−0.028	0.196	0.443
		$4C_{v,x} = 2.00$	0.159	−0.326	0.304	0.552
1	0.75	$C_{v,x} = 0.75$	−2.055	1.090	0.058 3	0.242
		$2C_{v,x} = 1.50$	−0.607	0.378	0.196	0.443
		$3C_{v,x} = 2.25$	−0.144	−0.044	0.358	0.599
		$4C_{v,x} = 3.00$	0.088	−0.343	0.513	0.717
1	1.00	$C_{v,x} = 1.00$	−2.105	1.087	0.098 2	0.314
		$2C_{v,x} = 2.00$	−0.675	0.364	0.304	0.552
		$3C_{v,x} = 3.00$	−0.221	−0.056	0.513	0.717
		$4C_{v,x} = 4.00$	0	−0.346	0.693	0.834
1	1.20	$C_{v,x} = 1.20$	−2.148	1.080	0.135 5	0.368
		$2C_{v,x} = 2.40$	−0.738	0.360	0.389	0.624
		$3C_{v,x} = 3.60$	−0.291	−0.057	0.624	0.791
		$4C_{v,x} = 4.80$	−0.068	−0.342	0.816	0.904

The literature [114, 117] also quotes explicit expressions for parameters $\bar{y}$, σ_y^2 and x_0:

$$\bar{y} = \ln\sigma_x - \ln|c| - \tfrac{1}{2}\ln(1 + c^2)\,, \tag{4.10}$$

$$\sigma_y^2 = \ln(1 + c^2)\,, \tag{4.11}$$

$$x_0 = \bar{x} - \frac{1}{c}\sigma_x\,, \tag{4.12}$$

TABLE 7.
Statistical processing of the sets of samples of a 10 000-element random sequence with triparametric log-normal distribution

Input and output parameters of a 10 000-element random sequence			
Inputs:		Outputs:	
$\bar{y} = 0.360$		$\bar{y} = 0.359$	
$\sigma_y = 0.624$		$\sigma_y = 0.626$	
$C_{s,y} = 0$		$C_{s,y} = 0$	
$\bar{x} = 1.00$		$\bar{x} = 1.006$	
$C_{v,x} = 1.20$		$C_{v,x} = 1.210$	
$C_{s,x} = 2.40$		$C_{s,x} = 2.317$	
$x_0 = -0.738$		$x_0 = -0.623$	
Characteristics of 500 samples			
	$n = 20$	$n = 25$	$n = 30$
$E(\bar{x})$	1.006 8	1.005 2	1.004 4
$\sigma(\bar{x})$	0.281 2	0.250 9	0.226 2
$C_v(x)$	0.279 3	0.249 6	0.225 2
$C_s(\bar{x})$	0.507 2	0.547 6	0.385 5
$\bar{x}_{max}$	2.139 2	2.120 1	1.911 0
$\bar{x}_{97.5}$	1.620 6	1.516 1	1.450 5
$\bar{x}_{2.5}$	0.520 1	0.566 7	0.605 9
$\bar{x}_{min}$	0.284 0	0.406 8	0.423 9
$E(C_v)$	1.161 4	1.169 3	1.170 1
$\sigma(C_v)$	0.277 3	0.236 8	0.221 9
$C_v(C_v)$	0.238 8	0.202 5	0.189 6
$C_s(C_v)$	1.226 1	0.679 2	0.763 8
$C_{v,max}$	2.717 6	2.092 4	2.057 5
$C_{v,97.5}$	1.785 7	1.697 6	1.684 1
$C_{v,2.5}$	0.784 5	0.791 3	0.819 0
$C_{v,min}$	0.546 0	0.686 1	0.700 0
$E(C_s)$	1.274 1	1.403 8	1.477 9
$\sigma(C_s)$	0.650 0	0.695 5	0.699 9
$C_v(C_s)$	0.510 1	0.495 4	0.473 5
$C_s(C_s)$	0.611 2	0.716 4	0.847 7
$C_{s,max}$	3.470 7	3.672 7	4.086 9
$C_{s,97.5}$	2.635 9	3.070 2	3.160 9
$C_{s,2.5}$	0.212 3	0.334 1	0.451 6
$C_{s,min}$	−0.170 6	−0.106 5	0.066 0

where

$$C_{s,x} = c^3 + 3c\,. \tag{4.13}$$

The unknown parameters $\bar{y}$, σ_y and x_0 were computed from equations (4.3), (4.4) and (4.9) in 16 variants for the following range of inputs: $\bar{x} = 1$, $C_{v,x} = 0.50–1.20$, $C_{s,x} = C_{v,x}$, $C_{s,x} = 2C_{v,x}$, $C_{s,x} = 3C_{v,x}$ and $C_s = 4C_{v,x}$. For the respective survey the reader is referred to Table 6.

When then parameters of the logarithms had been found, random sequences of 10 000 terms were modelled as absolutely random sequences with normal distribution determined by parameters $\bar{y}$ and σ_y. Reducing the modelled variables y to numbers gave us random variables x with log-normal distribution. All the pairs or random sequences were duly checked for the agreement between the input and the output parameters.

From the modelled 10 000-term sequences 500 random samples were produced; in the first period of our research [75] the lengths of the samples equalled 20, 25 and 30 terms, later [82], sets of samples with 40, 50 and 60 terms were added, which satisfactorily covered the lengths of the series that are most often available in hydrological practice. For each sample all the basic characteristics were computed, and the moment characteristics were then calculated for their 500-term sets (always for the given n); moreover, the lines of transgression of the sets of the characteristics were also plotted, from which the required two-sided confidence intervals (with the usual degree of significance of 5 percent) were easily derived. This detailed statistical procedure enabled us to express the complex probability properties of the samples and to derive their random and systematic errors.

For an example of statistical processing of the sets of samples of a single random sequence see Table 7, where the lengths of the samples have been limited to $n = 20$, 25 and 30 terms.

An example of the fluctuation of the characteristics within the bounds of the lines of transgression is shown in Fig. 17, where the lengths of the samples have again been cut to $n = 20$, 25 and 30 terms.

Of all the properties of the characteristics examined, the skewness of these characteristics is the most interesting. With the sample coeffcients of asymmetry, skewness increases in proportion to the growing length of the sample; with the sample means and the coefficients of variation, it is very often the reverse. The fluctuation of all the characteristics (measured by their coefficients of variation) decreases with the length of sample increasing, which has a positive effect upon the construction of the confidence interval and the dependability of the estimation.

The most important are the properties of the characteristics in their extreme values, which give expression to possible fluctuations of the random errors, and in their expected values, which are an expression of the possible fluctuations of the systematic errors with respect to the values of the parameters.

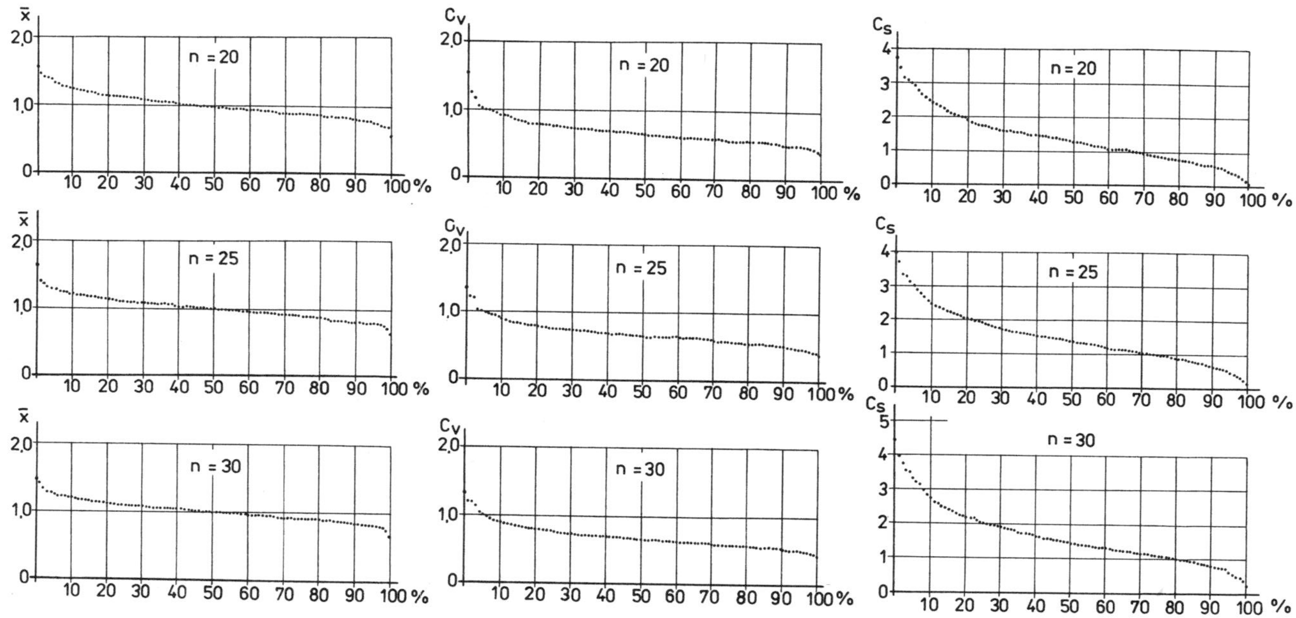

Fig. 17. Lines of transgression of 500 sample means, coefficients of variation and asymmetry, derived from a 10 000-element random sequence with triparametric log-normal distribution.

Inputs: $\bar{y} = -0.343$, $\sigma_y = 0.717$, $C_{s,y} = 0$, $\bar{x} = 1.00$,
$C_{v,x} = 0.75$, $C_{s,x} = 3.00$, $x_0 = 0.088$;

Outputs: $\bar{y} = -0.343$, $\sigma_y = 0.722$, $C_{s,y} = -0.058$,
$\bar{x} = 1.006$, $C_{v,x} = 0.747$, $C_{s,x} = 2.842$,
$x_0 = 0.138$.

The properties of the systematic errors examined manifested a clear tendency to form functional relationships with respect to the parameters, so that we were in a position to produce diagrams, similar to Rozhdestvenskiĭ's diagrams, enabling a rapid estimation of the unbiassed parameters (Figs 18 and 19).

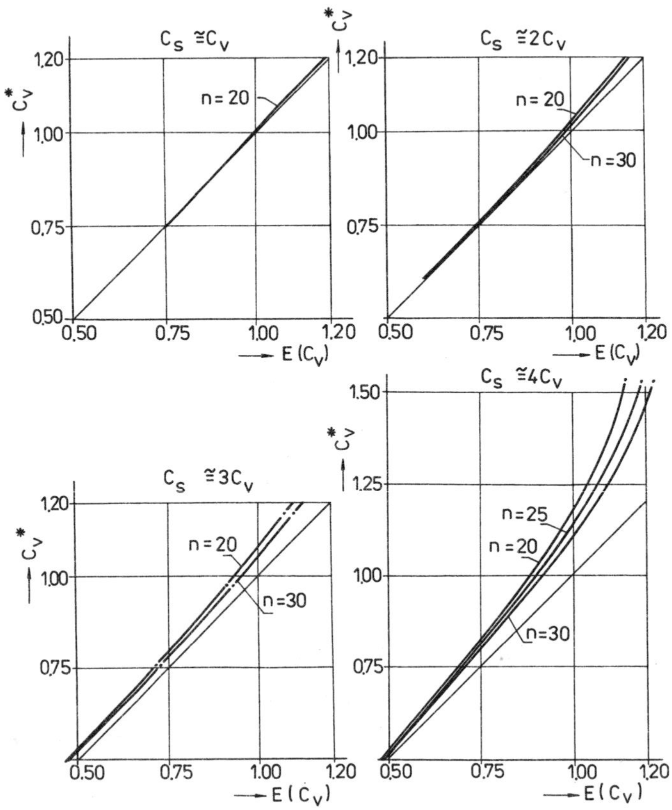

Fig. 18. Diagrams facilitating the estimation of the coefficients of variation for the log-normal distribution.

From the computation of the estimates of the parameters of sequences with log-normal distribution the following conclusions could be drawn:

1. Estimation of parameters based on a single sample can be burdened – as in the case of Pearson's distribution and gamma distribution – with considerable random errors, which grow with the length of the sample decreasing and the variability and asymmetry of the set increasing. Ways must therefore be sought of approximating to the unbiassed parameters with a methodologically substantiated estimator.

2. From the point of view of systematic errors, only the expected values of the sample means are unbiassed with respect to the parameters. The expected values of the sample coefficients of variation and asymmetry are one-sidedly biassed with respect to their long-term values (they are invariably lower than the parameters). With the coefficients of variation and asymmetry increasing, and

with the length of the sample decreasing, the systematic errors grow similarly as the random errors. The relatively greatest systematic errors can occur with the coefficient of asymmetry; the estimation of that parameter should therefore be given the closest attention.

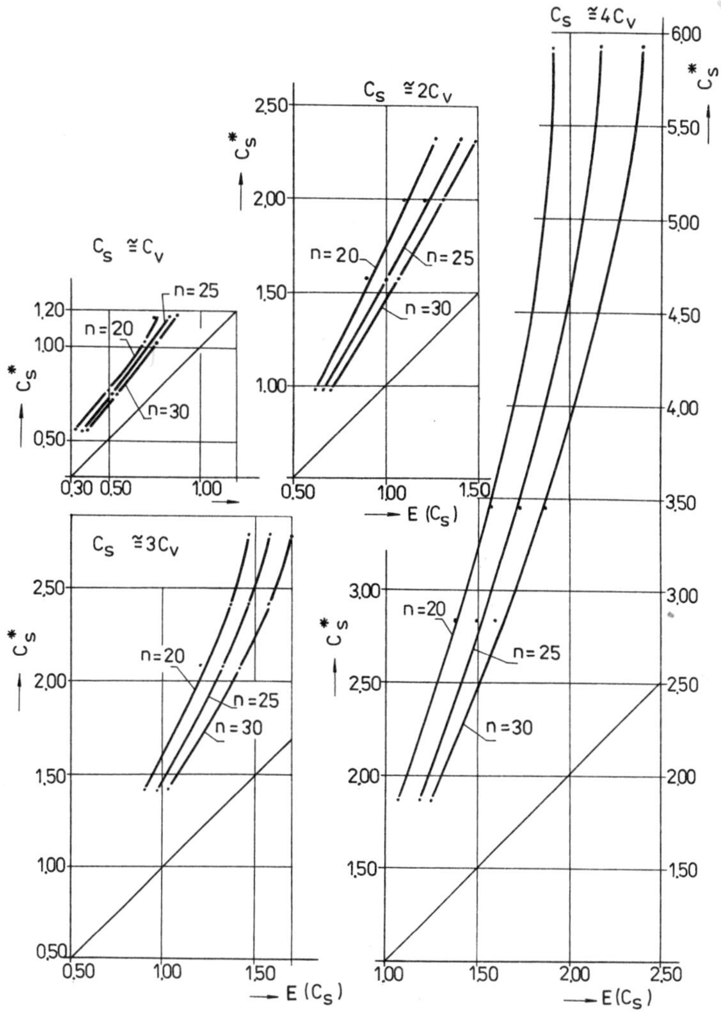

Fig. 19. Diagrams facilitating the estimation of the coefficients of asymmetry for the log-normal distribution.

3. From the properties of behaviour of the random and the systematic errors described it follows that engineering practice should draw a most important conclusion, which is that the design values of the parameters must be fully ascertained in keeping with the principles of the theory of estimation, and that,

if necessary, the biassed characteristics of the given sample should be corrected by the systematic errors being added to give unbiassed estimates of the parameters. And as far as the samples are concerned, the theory of estimation upholds the well-known requirement concerning their size (viz. measurement or observation should be as long as possible), in order that the possible random or systematic errors may be minimal.

Whereas the problem of estimating parameters in the biparametric case of log-normal distribution has satisfactorily been solved, the literature [92] gives attention to some difficulties involved in the search for the best properties of the parameters estimated in the case of triparametric distribution. Difficulties arise, for example, with the estimation of parameter x_0 localizing the position of the distribution. Problems are also posed by the selection of the estimation method itself, the success of which depends upon the skewness of the distribution. Since this type of distribution is preferred by, and frequently used in, hydrological practice, research should concentrate its efforts on solving the open problems in this sphere.

When the triparametric log-normal distribution was applied to flow series, our research showed that this type of distribution was a suitable model for the average daily flows, for example, and that it could often also be applied to average monthly flows. This type of distribution is however far less suitable for the determination of the properties of the average annual flows, where the logarithmic Pearson distribution of the IIIrd type proves to be the most suitable. The logarithmic Pearson distribution is highly adaptable, which gives it wide possibilites of application.

4.2.2 Estimation of parameters of a population with logarithmic Pearson distribution of the IIIrd type

Research into the methods of estimation of a similar scope was carried out for the logarithmic Pearson distribution of the IIIrd type. This type of distribution, linked with the method of moments, was recommended in 1967 by the American Water Resources Council as a standard method of analysis of the frequency of the occurrence of floods [11, 27]. The reason for this recommendation was the considerable adaptability of probability density, which can take on the most varied forms dependent upon the selection of three parameters. The logarithmic Pearson distribution therefore became widely applied by the theory of estimation of the unknown parameters of a population.

Although the Pearson distribution is currently being used in Czechoslovak water-engineering practice, its logarithmic modification has not been widely applied. In the literature available, particularly in the more recent numbers of some journals, we have however come across interesting contributions [11, 13,

27] dealing with the fundamental properties of the logarithmic Pearson distribution and the possibilities of its application to the processing of hydrological data. The article by Bobée [11] is particularly useful, analyzing in detail the probability density of the logarithmic Pearson distribution, its possible forms, and deriving the mutual relationships between the parameters of the logarithmic Pearson distribution and the parametrs of the Pearson distribution.

The logarithmic Pearson distribution is defined analogously to log-normal distribution. Variable x has logarithmic Pearson's distribution if its logarithm $y = \log_a x$ has Pearson's distribution. The logarithmic transformation of the given variables x can generally be considered to any arbitrary base a; either common or natural logarithms are generally used.

In constructing the model of random sequences with logarithmic Pearson's distribution the basic task was again to derive the input parameters of the logarithms of variables $y = \log_a x$ for the given sample of variables $x_1, x_2, \ldots,$ x_n. If sample x is given by characteristics only, the solution of the problem will prove rather difficult. The general equation of Pearson's density of the IIIrd type in the following form was Bobée's point of departure:

$$f(y) = \frac{|\alpha|}{\Gamma(\lambda)}[\alpha(y - m)]^{\lambda-1}\, e^{-\alpha(y-m)}, \tag{4.14}$$

where α, λ, m represent the parameters (with λ always positive).

Two cases can be distinguished here: if $\alpha > 0$, the skewness of the distribution is positive and $m \leqq y < +\infty$; if $\alpha < 0$, the skewness of the distribution is negative and $-\infty < y \leqq m$. In the special case of $m = 0$, we get the gamma distribution.

The moment characteristics are expressed as follows:

- the expected value:

$$\mu = m + \frac{\lambda}{\alpha}; \tag{4.15}$$

- the coefficient of variation:

$$C_v = \frac{\alpha}{|\alpha|}\frac{\lambda^{1/2}}{(\lambda + m\alpha)}; \tag{4.16}$$

- the coefficient of asymmetry:

$$C_s = \frac{\alpha}{|\alpha|}\frac{2}{\lambda^{1/2}} \tag{4.17}$$

Bobée derives the probability density, $g(x)$, of the logarithmic Pearson distribution from (4.14) in the following form:

$$g(x) = \frac{\alpha}{\Gamma(\lambda)}[\alpha(\log_a x - m)]^{\lambda-1} \exp^{\{-\alpha(\log_a x - m)\}} \frac{k}{x}, \tag{4.18}$$

where k denotes a constant dependent upon base a (on the assumption that $a > 1$) according to the following relationship:

$$k = \log_a e = (\ln a)^{-1}, \tag{4.19}$$

which can easily be expressed numerically.*)

In equation (4.18), α, λ, m are again parameters of the Pearson distribution.

If $\alpha > 0$, the Pearson distribution exhibits positive skewness, and the definition domain, D, of variable x in the logarithmic Pearson distribution is given by the following expression:

$$a^m = e^{m/k} \leqq x. \tag{4.20a}$$

If, on the contrary, $\alpha < 0$, the Pearson distribution exhibits negative skewness, and the values of variable x range within the following limits:

$$0 \leqq x \leqq a^m = e^{m/k}. \tag{4.20b}$$

The statistical parameters of the logarithmic Pearson distribution (the general moments of the order r about the origin) are obtained either with the help of the following expression:

$$(\mu_r') = \int_D x^r \, g(x) \, dx, \tag{4.21}$$

where $g(x)$ is the probability density (4.18), and D is the definition domain of variable x according to (4.20a, 4.20b), or with the expression:

$$(\mu_r') = \frac{e^{mr/k}}{(1 - r/\beta)^{\lambda}}, \tag{4.22}$$

where $\beta = \alpha k$.

The solving procedure:

---

*) If $a = e$, then $y = \ln x$ and $k = 1$. If $a = 10$, then $y = \log x$ and $k = 1/\ln 10 \doteq 0.434$.

For the sample set of values $(x_1, x_2, \ldots, x_n)$ the first three general moments are initially computed:*)

$$l_r = \frac{1}{n} \sum_{i=1}^{n} x_i^r \,. \tag{4.23}$$

Their numerical values are then compared with expressions (4.22), which comprise the unknown parameters λ, m, and β. This will result in equations for computing the unknown parameters of the Pearson distribution (viz. the parameters of logarithms of the given variables x_i). Thus, if we take the logarithm of expression (4.22) for the first three general moments, we get the following system of equations:

$$\log l_1 = m - \lambda \log \left[1 - (1/\beta)\right], \tag{4.24}$$

$$\log l_2 = 2m - \lambda \log \left[1 - (2/\beta)\right], \tag{4.25}$$

$$\log l_3 = 3m - \lambda \log \left[1 - (3/\beta)\right], \tag{4.26}$$

with the help of which the following unknown parameters are expressed:

$$\frac{\log \{[1 - (1/\beta)]^3/[1 - (3/\beta)]\}}{\log \{[1 - (1/\beta)]^2/[1 - (2/\beta)]\}} = \frac{\log l_3 - 3\log l_1}{\log l_2 - 2 \log l_1} = B \,, \tag{4.27}$$

$$\lambda = \frac{\log l_2 - 2 \log l_1}{\log \{[1 - (1/\beta)]^2/[1 - (2/\beta)]\}} \,, \tag{4.28}$$

$$m = \log l_1 + \lambda \log \left[1 - (1/\beta)\right]. \tag{4.29}$$

For various values of B the corresponding values of α are tabulated (see Table 8); the values of β are determined from the following relationship:

$$\beta = \frac{\alpha}{\ln 10}$$

The last step is the computation of the desired characteristics $\bar{y}$, $C_{v,y}$ and $C_{s,y}$ from the values of B, α, β, λ, m.

The procedure indicated differs from the procedure of the Water-Engineering Council in that the characteristics of the Pearson distribution are determined

*) If sample x_i is given by characteristics $\bar{x}$, C_v, C_s only, then the general moments (4.23) are derived from the known relationships between the central and the general moments.

TABLE 8. Relationship Between α and *B* (Common Logarithms)

α	*B*	α	*B*	α	*B*	α	*B*
6.910	23.720 4	28	3.207 54	−0.001	2.040 79	−29	2.867 35
6.912	21.527 5	29	3.198 57	−0.002	2.045 21	−30	2.871 06
6.914	20.205 5	30	3.190 35	−0.003	2.048 26	−31	2.874 58
6.916	19.258 5	31	3.182 78	−0.004	2.050 68	−32	2.877 91
6.918	18.521 5	32	3.175 79	−0.005	2.052 73	−33	2.881 06
6.920	17.918 9	33	3.169 32	−0.006	2.054 52	−34	2.884 06
6.922	17.409 7	34	3.163 30	−0.007	2.056 13	−35	2.886 91
6.924	16.969 3	35	3.157 70	−0.008	2.057 60	−36	2.889 62
6.926	16.581 5	36	3.152 48	−0.009	2.058 96	−37	2.892 21
6.928	16.235 3	37	3.147 58	−0.01	2.060 23	−38	2.894 68
6.93	15.923 0	38	3.143 00	−0.02	2.070 05	−39	2.897 03
6.94	14.703 9	39	3.138 68	−0.03	2.077 28	−40	2.899 29
6.95	13.830 9	40	3.134 63	−0.04	2.083 29	−41	2.901 44
6.96	13.154 4	41	3.130 80	−0.05	2.088 55	−42	2.903 51
6.97	12.604 5	42	3.127 18	−0.06	2.093 29	−43	2.905 49
6.98	12.142 8	43	3.123 76	−0.07	2.097 65	−44	2.907 39
6.99	11.746 1	44	3.120 52	−0.08	2.101 73	−45	2.909 22
7.00	11.399 2	45	3.117 44	−0.09	2.105 56	−46	2.910 98
7.01	11.091 6	46	3.114 52	−0.1	2.109 20	−47	2.912 67
7.02	10.815 9	47	3.111 74	−0.2	2.139 18	−48	2.914 29
7.03	10.566 5	48	3.109 09	−0.3	2.163 04	−50	2.917 37
7.04	10.339 1	50	3.104 15	−0.4	2.183 74	−52	2.920 23
7.05	10.130 4	52	3.099 64	−0.5	2.202 35	−54	2.922 90
7.06	9.937 9	54	3.095 51	−0.6	2.219 44	−56	2.925 40
7.07	9.759 4	56	3.091 70	−0.7	2.235 33	−58	2.927 74
7.08	9.593 17	58	3.088 19	−0.8	2.250 23	−60	2.929 94
7.09	9.437 80	60	3.084 93	−0.9	2.264 30	−62	2.932 01
7.10	9.292 07	62	3.081 91	−1	2.277 64	−64	2.933 96
7.11	9.154 98	64	3.079 10	−1.5	2.335 92	−66	2.935 81
7.12	9.025 65	66	3.076 47	−2.0	2.383 98	−68	2.937 55
7.13	8.903 35	68	3.074 01	−2.5	2.424 80	−70	2.939 20
7.14	8.787 42	70	3.071 71	−3.0	2.460 13	−72	2.940 76
7.15	8.677 31	72	3.069 54	−3.5	2.491 12	−74	2.942 25
7.2	8.198 63	74	3.067 50	−4.0	2.518 58	−76	2.943 66
7.3	7.488 02	76	3.065 58	−4.5	2.543 11	−78	2.945 01
7.4	6.975 94	78	3.063 76	−5.0	2.565 19	−80	2.946 29
7.5	6.583 78	80	3.062 04	−5.5	2.585 17	−82	2.947 51
7.6	6.271 07	82	3.060 41	−6.0	2.603 35	−84	2.948 68
7.7	6.014 35	84	3.058 87	−7.0	2.635 24	−86	2.949 80
7.8	5.798 91	86	3.057 40	−8.0	2.662 31	−88	2.050 87
8	5.455 73	88	3.056 00	−9	2.685 60	−90	2.951 90
9	4.551 07	90	3.054 67	−10	2.705 85	−92	2.952 88
10	4.145 23	92	3.053 41	−11	2.723 63	−94	2.953 83
11	3.910 92	94	3.052 20	−12	2.739 37	−96	2.954 74
12	3.757 41	96	3.051 04	−13	2.753 40	−98	2.955 61
13	3.648 72	98	3.049 93	−14	2.765 99	−100	2.956 45
14	3.567 59	100	3.048 87	−15	2.777 35	−125	2.964 78

TABLE 8. *(continued)*

α	B	α	B	α	B	α	B
15	3.504 65	125	3.038 62	−16	2.787 66	−150	2.970 43
16	3.454 36	150	3.031 93	−17	2.797 05	−175	2.974 52
17	3.413 25	175	3.027 21	−18	2.805 64	−200	2.977 62
18	3.379 00	200	3.023 71	−19	2.813 53	−250	2.981 99
19	3.350 02	250	3.018 86	−20	2.820 81	−300	2.984 94
20	3.325 17	300	3.015 65	−21	2.827 53	−350	2.987 06
21	3.303 63	400	3.011 68	−22	2.833 77	−400	2.988 65
22	3.284 77	500	3.009 32	−23	2.839 58	−500	2.990 89
23	3.268 13	600	3.007 75	−24	2.844 99	−600	2.992 40
24	3.253 33	700	3.006 63	−25	2.850 04	−700	2.993 47
25	3.240 08	800	3.005 80	−26	2.854 78	−800	2.994 28
26	3.228 16	900	3.005 15	−27	2.859 22	−900	2.994 92
27	3.217 36	1 000	3.004 63	−28	2.863 41	−1000	2.995 42

TABLE 9. Input parameters of models with logarithmic Pearson's IIIrd type distribution

Paramaters of the given variables x			Parameters of the logarithms of variables x		
$\bar{x}$	$C_{v,x}$	$C_{s,x}$	$\bar{y}$	$C_{v,y}$	$C_{s,y}$
1	0.50	$C_{v,x} = 0.50$	−0.066	−3.958	−0.947
		$2C_{v,x} = 1.00$	−0.056	−4.093	−0.449
		$3C_{v,x} = 1.50$	−0.050	−4.198	−0.078
		$4C_{v,x} = 2.00$	−0.045	−4.306	0.208
1	0.75	$C_{v,x} = 0.75$	−0.182	−2.657	−1.348
		$2C_{v,x} = 1.50$	−0.128	−2.815	−0.607
		$3C_{v,x} = 2.25$	−0.105	−2.936	−0.174
		$4C_{v,x} = 3.00$	−0.092	−3.032	0.113
1	1.00	$C_{v,x} = 1.00$	−0.417	−2.054	−1.692
		$2C_{v,x} = 2.00$	−0.228	−2.201	−0.726
		$3C_{v,x} = 3.00$	−0.175	−2.318	−0.270
		$4C_{v,x} = 4.00$	−0.141*)	−2.402*)	0
1	1.20	$C_{v,x} = 1.20$	−0.767	−1.790	−1.942
		$2C_{v,x} = 2.40$	−0.326	−1.908	−0.798
		$3C_{v,x} = 3.60$	−0.240	−2.015	−0.339
		$4C_{v,x} = 4.80$	−0.204	−2.091	−0.081

*) Since in this case the solution is not defined, the parameters were found with the help of a statistical experiment using several models.

from the moments of the given variables x_i, without their logarithms being computed. This will ensure a closer fit of the distribution in the extremes.

Since the computation of characteristics $\bar{y}$, $C_{v,y}$, $C_{s,y}$ from the given characteristics $\bar{x}$, $C_{v,x}$, $C_{s,x}$ is both numerically exacting as well as highly time-consuming (and moreover fairly sensitive to the accuracy of the computations of the auxiliary values), it should be programmed for a computer. In our research we studied the process of estimation using sixteen models, the inputs of which are listed in Table 9.

The adaptation of the moments method to suit the models with logarithmic Pearson's distribution involved the same steps as those required by the adaptation of that method to the models with log-normal distribution. 10 000-term random sequences were modelled for sixteen variants of inputs to give 500 samples of the lengths of 20, 30, and 40 terms. As with the models with log-normal distribution, diagrams were plotted from the results of the investigation of the law-governed behaviour of the characteristics in the individual variants, showing schematically the relationships between the biassed and the unbiassed estimates of the coefficients of variation and asymmetry (see Figs 20 and 21). The diagram can again be used for ready determination of the unbiassed estimates of these coefficients.

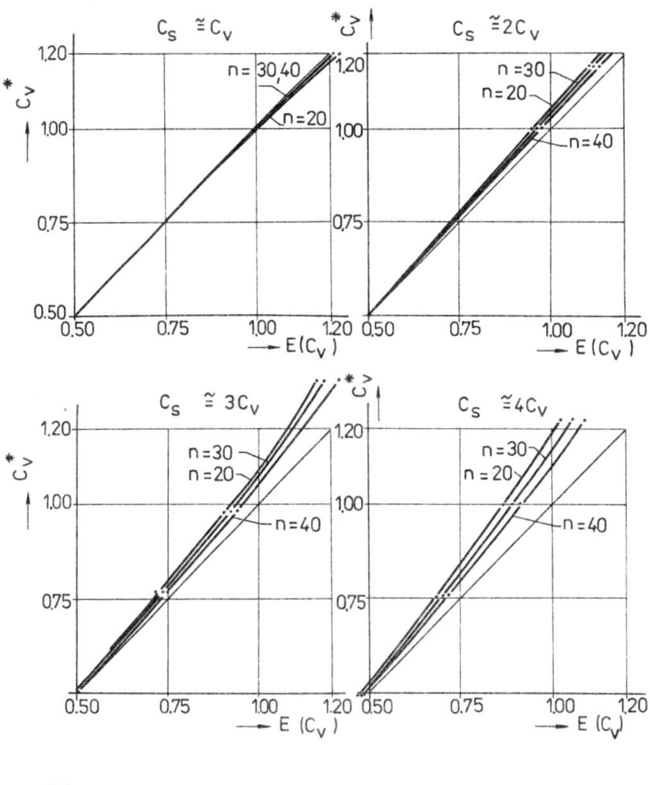

Fig. 20. Diagrams for the estimation of the coefficients of variation for the logarithmic Pearson distribution.

Studying the properties of the characteristics, their random and their systematic errors, we arrived at conclusions similar to those for the sequences with log-normal distribution. The fluctuation of all the characteristics decreases with

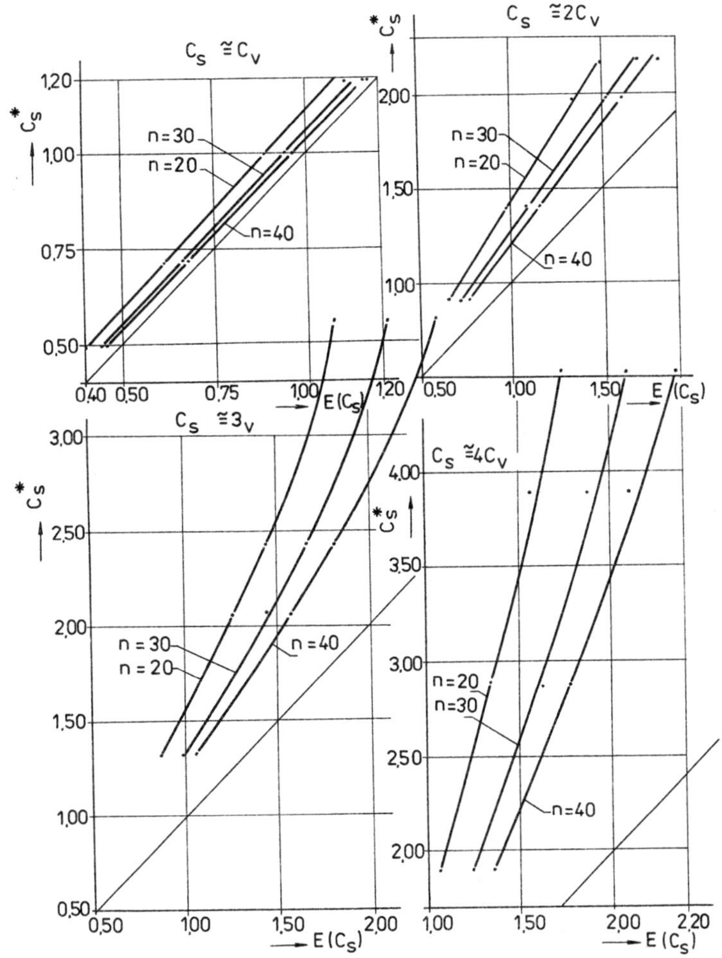

Fig. 21. Diagrams for the estimation of the coefficients of asymmetry for the logarithmic Pearson distribution.

the length of the samples increasing: a longer sample thus positively affects the dependability of an estimate, which is both logical and well-known.

An analysis of the behaviour of the systematic errors did not bring any further surprises. Only the expected values of the sample means were unbiassed; the expected values of the sample coefficients of variation and assymmetry were one-sidedly biassed (viz. again lower as compared with the parameters); and the highest attention must be given to the estimation of the coefficient of asymmetry.

More interest was aroused by the results of the comparison of the paired confidence intervals of the sequences with log-normal and logarithmic Pearson's distributions (for the same inputs). This analysis has been described in more detail in [75]. From the ninety-six values of confidence intervals compaired only twenty-eight (i. e. 29 percent) were larger with the logarithmic Pearson distribution. This indicates that the application of the logarithmic Pearson distribution to hydrological series is suitable from the point of view of both the basic property of that distribution, i. e. close fit of probability density, and the lesser random errors of the sample characteristics.

The comparison of the systematic errors in the estimation of parameters in the given two types of distribution showed certain tendencies dependent upon the C_s/C_v ratio. With the lower values of that ratio, the systematic errors were invariably less pronounced in the models with logarithmic Pearson's distribution; with the higher values of that ratio, the errors were markedly minor in the models with log-normal distribution.

The properties exhibited by the estimators of the parameters of sequences with logarithmic Pearson's distribution are thus on the whole positive, which ensures a wide practical application for this type of distribution. And since the logarithmic Pearson distribution is highly adaptable, it can be used for the solution of a wide range of problems.

4.3 Mutual relationships between the random, probable and systematic errors of parameter estimation

The investigation of the mutual relationships between the random, probable and systematic errors was prompted by both the markedly different properties of the individual types of errors and the mostly simplified ideas concerning the sufficient assessment of the dependability of hydrological computations on the assumption of the fluctuation of hydrological variables within the bounds of probable errors only, or the standard deviations of the individual sample characteristics. Systematic errors are often neglected and their underevaluation involves high risks concerning the reliability of estimation, particularly as far as the shorter samples and the more extreme distributions of probability are concerned.

In the Soviet water-engineering literature [96] considerable attention is paid to the investigation of the properties of sample characteristics. Under the assumption of Pearson's distribution of the IIIrd type, the following approximate expressions are quoted for their standard deviations:

– for the standard deviation of standard deviations,

$$\sigma(s) = \frac{\sigma}{\sqrt{2n}} \sqrt{1 + 3\,C_v^2}\,; \tag{4.30}$$

and with the correlation of the neighbouring terms of the series,

$$\sigma(s) = \frac{\sigma}{\sqrt{2n}} \sqrt{(1 + 3C_v^2)\left(1 + \frac{r}{1 + r}\right)}; \tag{4.31}$$

– for the standard deviation of the coefficients of variation,

$$\sigma(C_v) = \frac{C_v}{\sqrt{2n}} \sqrt{1 + C_v^2}; \tag{4.32}$$

– for the standard deviation of the coefficients of asymmetry,

$$\sigma(C_s) = \sqrt{\frac{6}{n}} \sqrt{1 + 6C_v^2 + 5C_v^4}; \tag{4.33}$$

and the shortened approximate (empirical) expression,

$$\sigma(C_s) = \sqrt{\frac{6}{n}(1 + C_v^2)}; \tag{4.34}$$

– for the standard deviation of the ratio C_s/C_v,

$$\sigma\left(\frac{C_s}{C_v}\right) = \frac{1}{C_v\sqrt{n}} \sqrt{6 + 26C_v^2 + 24C_v^4}. \tag{4.35}$$

The evaluation of the dependability of the expressions quoted above revealed relatively good agreement with the values of the standard deviations observed in real series for the values of $C_v < 1$. The same literature [96] therefore also quotes Blokhinov's empirical expressions suitable for the higher values of C_v:

$$\sigma(C_v) = \frac{n}{n + 4C_v^2} \frac{C_v}{\sqrt{2n}} \sqrt{1 + C_v^2}, \tag{4.36}$$

and with the correlation of the neighbouring terms of the series,

$$\sigma(C_v) = \frac{n}{n + 4C_v^2} \frac{C_v}{\sqrt{2n}} \sqrt{(1 + C_v^2)\left(1 + \frac{3C_v r^2}{1 + r}\right)}. \tag{4.37}$$

For the standard deviation of the sample coefficients of variation, the Czechoslovak literature [30] quotes an expression similar to equation (4.32), and for the standard deviation of the sample coefficients of asymmetry the expression is identical with (4.33).

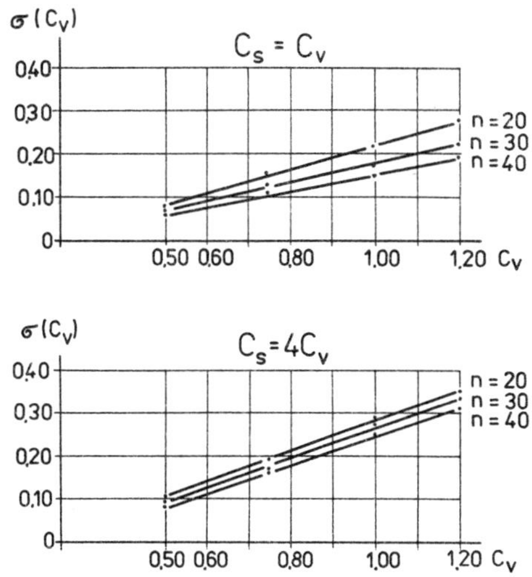

Fig. 22. Standard deviations of sample coefficients of variation from Pearson's IIIrd type distribution.

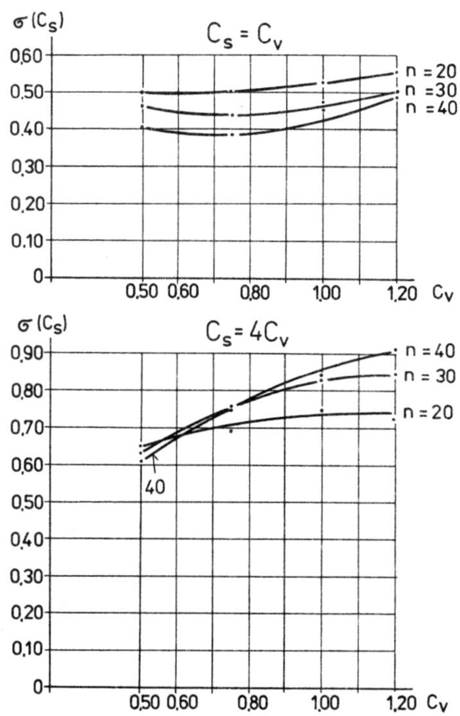

Fig. 23. Standard deviations of sample coefficients of asymmetry from Pearson's IIIrd type distribution.

We first concentrated upon the comparison of the magnitudes of the standard deviations of the sample coefficients of variation and asymmetry from Pearson's distribution of the IIIrd type, which were both computed using expressions (4.32) and 4.33), and were derived from the sets of five hundred random samples of modelled 10 000-term sequences. This analysis consisted in the characteristics of all the random samples having been computed and their 500-element sets then further processed.

The results are shown in Figs 22 and 23. The dependence of the standard deviations of the sample coefficients of variation upon the coefficient of variation of the universe, derived from the sets of random samples, as presented in Fig. 22, confirmed good agreement with the values of the standard deviations according to expression (4.32) for $C_s = C_v$. For larger C_s/C_v ratios the values of the standard deviations are slightly higher as far as the solution based upon random samples is concerned. The differences are however negligible, and for the purposes of orientation use can be made of expression (4.32).

An interesting curve is shaped by the dependence of the standard deviations of the sample coefficients of asymmetry upon the coefficient of variation of the population (Fig. 23). This relationship was again derived from the sets of random samples. With the parameter growing, the standard deviations increase, similarly to $\sigma(C_v)$. The effect of the length of the sample, n, however, differs with different values of the C_s/C_v ratio. And the important fact should be noted that expression (4.33) gives substantially higher values as compared with our solution. Using it will therefore raise the fluctuation of the sample coefficients of asymmetry, which can give rise to a misleading idea of the possible fluctuation of these coefficients being expressed more safely. In the assessment of the admissible fluctuation of the sample characteristics, that assumption however leads to wider, and thus much less sure, confidence intervals.

On the basis of the sets of 500 random samples of 10 000-element random sequences, which we modelled for the pre-selected parameters, we also examined the relationships between the probable errors and the widths of 95 percent confidence intervals of the characteristics [76].

For the models of random sequences we chose three of the most frequently applied types of distribution: Pearson's IIIrd type, triparametric log-normal, and logarithmic Pearson's IIIrd type.

For each of these distributions we constructed sixteen models with the following inputs: coefficients of variation $C_v = 0.50, 0.75, 1.00$ and 1.20; coefficients of asymmetry $C_s = C_v$, $C_s = 2C_v$, $C_s = 3C_v$ and $C_s = 4C_v$; the means were always chosen as equal to unity.

For each of the characteristics (sample means, coefficients of variation and asymmetry) the relationships were examined on a set of forty-eight values (three lengths of samples for each of the sixteen models, viz. 20, 30 and 40-term lengths with Pearson's distribution and logarithmic Pearson's distribution, and 20, 25 and 30-term lengths with log-normal distribution).

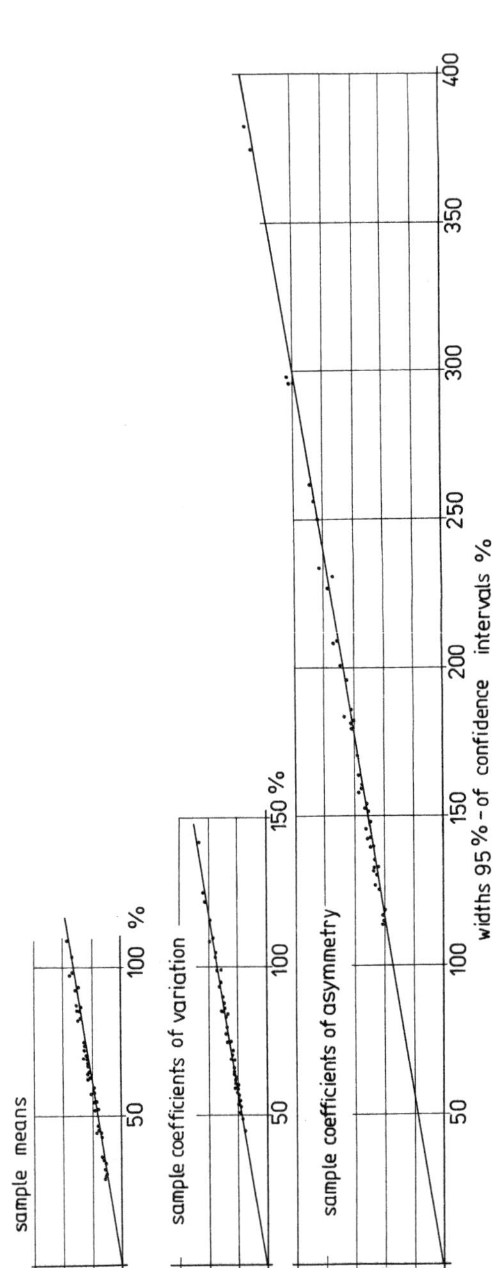

Fig. 24. Relationship between probable errors and the widths of 95 % confidence intervals of sample characteristics of random sequences with Pearson's IIIrd type distribution.

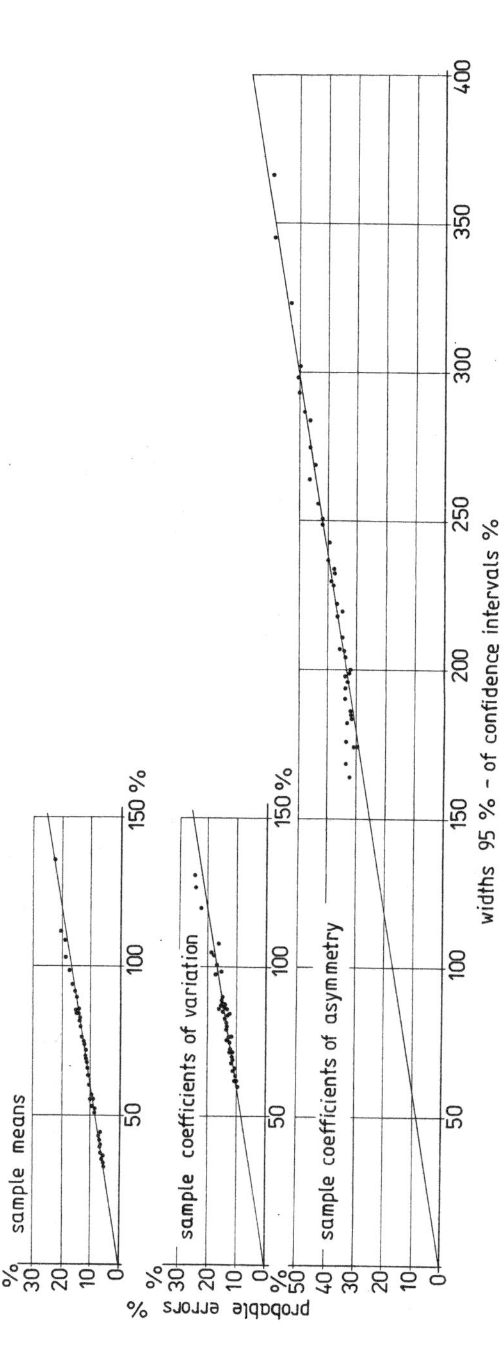

Fig. 25. Relationship between probable errors and the widths of 95 % confidence intervals of sample characteristics of random sequences with logarithmic Pearson IIIrd type distribution.

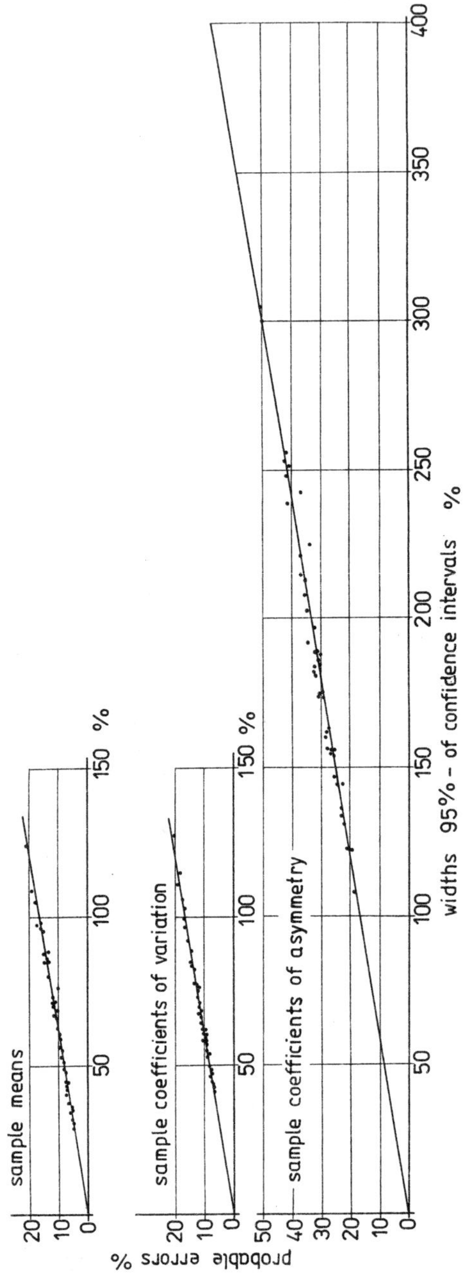

Fig. 26. Relationship between probable errors and the widths of 95 % confidence intervals of sample characteristics of random sequences with triparametric log-normal distribution.

The probable error was considered in the light of its usual meaning derived from the properties of the density of normal distribution. The $\pm$ 0.674 5 σ deviations from the mean enclose half the area limited by density; any arbitrary characteristic can thus lie within these bounds and outside, with a 50 percent probability. The following value,

$$d(u) = 0.674\,5\ \sigma(u) \doteq \tfrac{2}{3}\,\sigma(u) \tag{4.38}$$

is called the probable error, which can be viewed (like standard deviation) as a measure of variance of an arbitrarily chosen characteristic u.

The graphical representations of the results of this examination are given in Figs 24 to 26. It is noteworthy that in all the cases examined the relationships of correlation obtained were very close, and of linear character, with which the widths of 95 percent of the confidence intervals reach an average of six times the probable errors (or four times the standard deviations). This result can be explained by the properties of probability densities: with Gaussian distribution, the interval $(E(u) \pm 2\sigma(u))$ encloses 95.45 percent of its area; and an approximately similar relationship also holds for distributions with mild skewness, or for several cases of asymmetric distribution (e. g. Pearson's distribution).

The results of our examination show that neither the probable error nor the standard deviation suffices to reliably express the fluctuation of sample characteristics, which fluctuate within wide bounds suitably expressible by confidence intervals.

The relationships between the probable and the systematic errors are much more complex. We examined them in the same sixteen models of random sequences with log-normal and logarithmic Pearson's distributions related to the values of the two parameters, C_v and C_s, their ratio, C_s/C_v, and the length of the sample, n.

A characteristic example of these relationships, formulated for log-normal distribution is presented in Fig. 27. The left part of the diagram shows the plots of the probable and the systematic errors in the estimation of the coefficient of variation. Up to the value of $C_v = 1$ the probable errors are greater for all the ratios C_s/C_v examined. With $C_v = 1.20$ and $C_s/C_v = 4$, however, it is the systematic errors that are greater, and disregarding them may lead to grave errors in water-engineering computations.

The other three columns of the diagram show plots of the probable and the systematic errors in the estimation of the coefficient of asymmetry. The differences between these errors grow markedly with increasing skewness. With the highest value examined, $C_s = 4.80$ and $C_s/C_v = 4$, the systematic errors are already multiples of the probable errors, whereas with $C_s/C_v = 1$ the differences between these errors are minimal. This property can be accounted for by the fact that the fluctuation of the sample coefficients of asymmetry measured by their coefficient of variation (or their relative probable error) varies quite moderately with the skewness of the universe increasing. The systematic errors, however, are extremely sensitive to the rise of C_s.

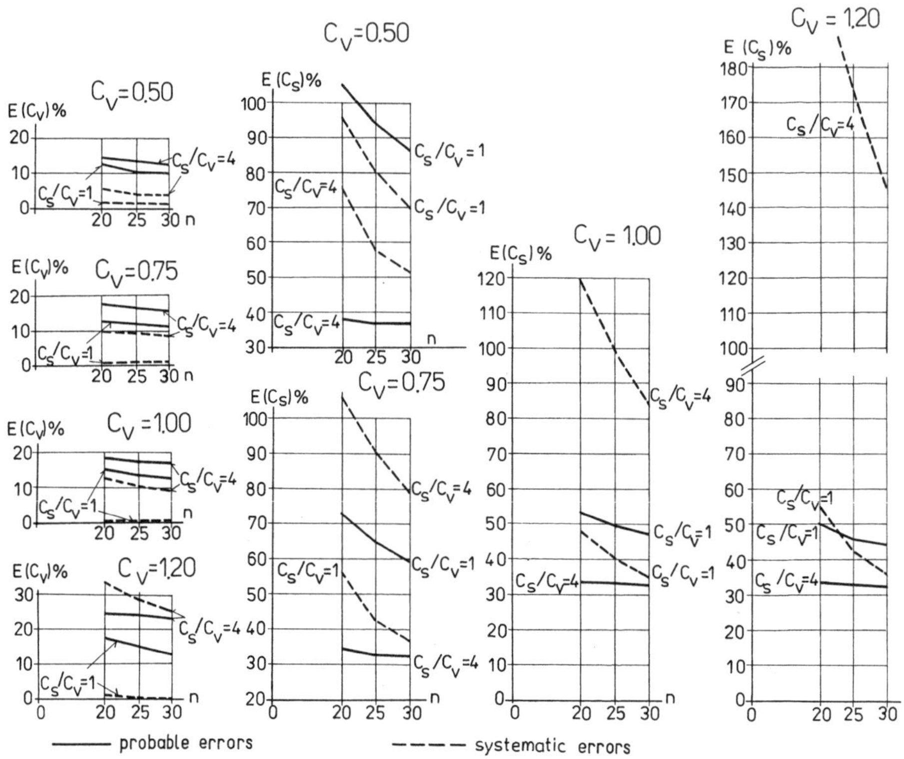

Fig. 27. Log-normal distribution: comparison of the probable and the systematic errors.

These results led to the conclusion that with the series of random variables it was essential to give attention, besides their variation (which is most conveniently expressed by the confidence intervals), to their systematic errors, and to take due account of these errors wherever unbiassed parameters are to be estimated. Systematic errors should be given special attention with the more elevated values of C_v and C_s, and ratio C_s/C_v. In water engineering these cases are of extreme importance, particularly when the floods of the more limited drainage basins are to be dealt with, exhibiting considerable variation and skewness of the maximum annual flows.

4.4 Effect of extreme sample elements on parameter estimation

The effect of the extreme values of some of the terms of random sequences on the magnitude of the bias of the sample characteristics, and thus also on the estimation of parameters, had attracted the interest of the water engineers even before the methods of the theory of estimation were fully elaborated. The experience gained indicated that a single extreme term of a sequence (e. g. a year with extreme flood flows) could markedly affect the sample characteristics, particularly the moments of higher order, which could considerably differ from the long term parameters. Therefore, approximative procedures should be sought that could at least reduce the intensity of this negative effect and come closer to justified estimates of parameters.

The simplest and the oldest method, widely used, is that of correcting the maximum values to lower values (e. g. average values), or excluding completely the maximum values that have occurred. This procedure was most often substantiated by the assertion that the probability of the extreme occurring was very low and out of proportion to the length of the observation. It is evident that this line of reasoning is hardly acceptable in view of contemporary knowledge of the relationships between the random samples and the universe, because each term of a real sequence ascertained by observation or measurement is an element of its sample and, simultaneously, of a larger, though unknown, population. No conclusions concerning the parameters of the population that the extreme ascertained belongs to can thus be drawn from samples with the extreme value reduced. And besides, the real probability of occurrence can hardly be ascertained with the help of an analysis of the given sample only: the ascertainment of that probability must be based upon the distribution of the population, which is however the subject of estimation.

In Section 2.2 we mentioned another simple method of estimating parameters based upon a comparative analysis of the properties of the sample and the properties of an analogue. This method did not necessitate any reduction of the extreme terms but the biassed characteristic itself, suspected of having been biassed by an extreme phenomenon. This characteristic was first compared with the characteristic of the analogue within the same period of time, and, additionally, its time-related (non-stationary) development and relationship to the parameter was monitored. The biassed characteristic was then corrected in accordance with that relationship and the analogue. But uncertainty arose if the correlation between the given series and the analogue proved less satisfactory.

Defining the effect of the extreme elements of a sample upon parameter estimation is complicated by two circumstances; firstly, the measure of the bias caused by the extreme; and secondly, the estimation of the parameters itself, which should give more scope to the random errors of the respective characteristics dependent upon the given extreme than to systematic errors. The second point proves particularly complex, for it requires that the representativeness of the given sample, as well as the relationship of that sample to the population, should be assessed and the random errors estimated.

As in the tasks discussed above, modelling methods proved to be the most suitable for the solution of this problem. In our research [81] we studied the behaviour of the extremes in 500 random (30-element) samples generated from 10 000-term modelled sequences.

The significance of the extreme in each sample was assessed using the non-parametric Dixon test and simultaneously, for purposes of comparison, also the parametric Grubbs test. This enabled us to classify the samples and their characteristics into three groups:

1. the characteristics biassed by the extremes at a significance level of 5 percent;
2. the characteristics biassed by the extremes at a significance level of 1 percent;
3. the characteristics unbiassed by the extremes (i. e. the extremes are statistically insignificant).

The characteristics were arranged according to magnitude into transgression lines of 500 values, and simultaneously classified into one of the three groups. This made it possible to assess the behaviour and the effect of the extremes on the characteristics, as well as the possibilities of their being corrected to give unbiassed parameters.

A most important aspect of the methodological approach was the application of Dixon's and Grubbs's tests to the assessment of the significance of the extremes. With the Dixon test the procedure is as follows [95]:
the test criterion is formulated by the following expression,

$$Q_n = \frac{x_n - x_{n-1}}{x_n - x_1}, \tag{4.39}$$

where x_n is the tested extreme (in the given case the highest) value in the sample, and x_1 is the lowest value in the sample.

Q_n is assumed to have Gaussian distribution. This assumption was verified by the respective characteristics being computed. In all cases the sets of variables Q_n were shown to exhibit negligible (near zero) skewness, and their distribution could therefore be regarded as approximately normal.

As the next step, the value of Q_n computed was compared with the critical value tabulated [95]. For $Q_n \geqq Q_{n,p}$ the null hypothesis, i.e. that x_n was an extreme (maximum) value, was adopted at the respective level of significance (1 percent or 5 percent).

The application of the Grubbs test involves an analogous procedure. The testing criterion is formulated by the following expresion:

$$T_n = \frac{x_n - \bar{x}}{s_x}, \tag{4.40}$$

where x_n is again the highest value in the sample, and $\bar{x}$, s_x are sample characteristics (the mean, the standard deviation). It is again assumed that variable T_n has Gaussian distribution. As in the Dixon test, the verification of this assumption gave positive results.

As the next step, the value of T_n computed is compared with the critical value [95]. For $T_n \geqq T_{n,p}$ the null hypothesis, i.e. that the value of x_n tested was an extreme value, is then adopted at the respective level of significance (1 percent or 5 percent).

Analogously with the verification of the significance of maxima, minima can also be verified as far as their significance is concerned. In this case expression (4.39) is given the following form:

$$Q_n = \frac{x_2 - x_1}{x_n - x_1}, \tag{4.41}$$

and expression (4.40) is changed to

$$T_n = \frac{\bar{x} - x_1}{s_x}, \tag{4.42}$$

where x_1 stands for the lowest value in the sample, x_2 for the second (last but one) lowest value in the sample, and $\bar{x}$, s_x again for the sample characteristics.

In our case, the minima nearly always proved to be insignificant, so that our attention could be given solely to the maxima.

And further analyses showed that the maxima can affect the values of the characteristics in various ways, which makes correcting them to unbiassed parameters more difficult. Two problems can arise in practice in this context, as follows.

1. The extreme element need not always lead to an extreme value of the characteristic. This can be explained by the occurrence of the individual values of the sample (e. g. the sample may contain a significant extreme, the other elements can however be relatively uniform). The characteristic can even acquire values lying below the average, so that the systematic error will hardly suffice to correct it to a parameter.

2. Remarkable properties are exhibited by the characteristics arranged into lines of transgression and identified according to their statistical significance. In

spite of the fact that a number of models tend prevailingly to group together characteristics of equal significance (for instance, the values statistically significant at a certain level gravitate towards the left part of the line of transgression, i. e. the zone of lower probabilities), characteristics of the same group can also occur in the higher probabilities transgression zone. The occurrence in the transgression lines of characteristics with the extremes identified can thus be wholly accidental and the tendencies group can be less embracing. It follows that

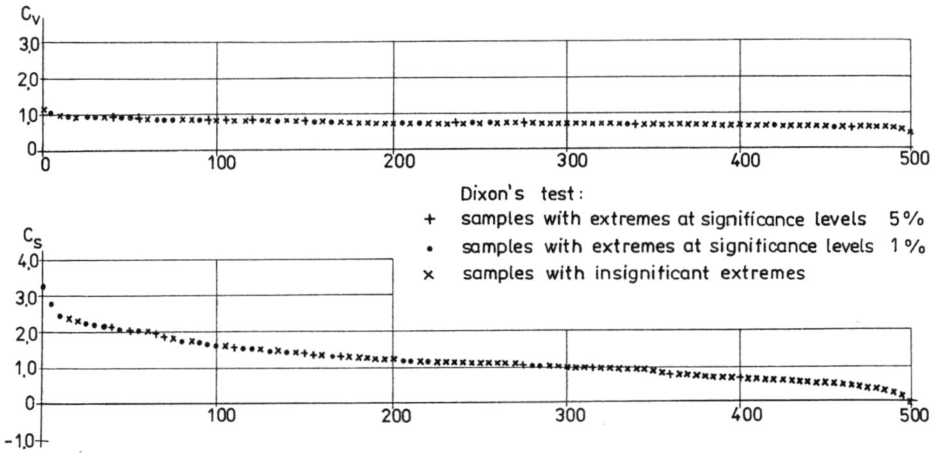

Fig. 28. Lines of transgression of identified sample coefficients of variation and asymmetry in a 10 000-element random sequence with Pearson's IIIrd type distribution.
Inputs: $\bar{Q} = 1.00$, $C_v = 0.75$, $C_s = 1.50$;
Outputs: $\bar{Q} = 1.013\,8$, $C_v = 0.760\,6$, $C_s = 1.593\,9$.

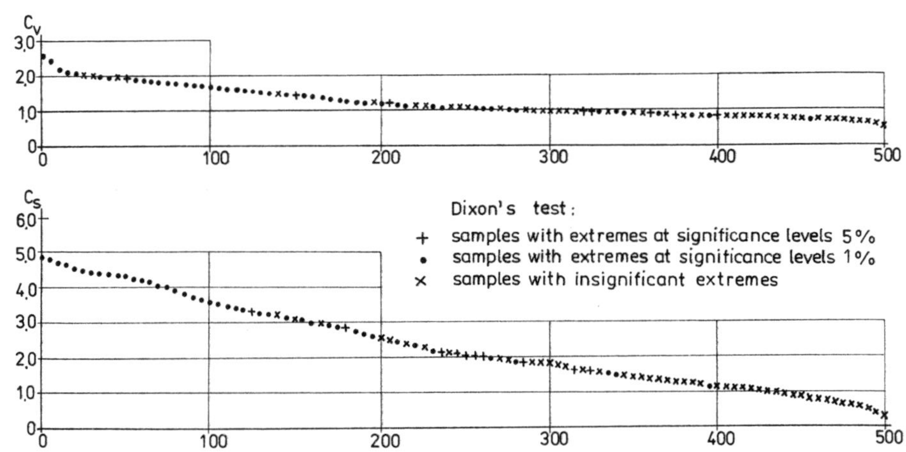

Fig. 29. Lines of transgression of identified sample coefficients of variation and asymmetry in a 10 000-element random sequence with log-normal distribution
Inputs: $\bar{Q} = 1.00$, $C_v = 1.50$, $C_s = 6.00$;
Outputs: $\bar{Q} = 1.056\,6$, $C_v = 1.548\,5$, $C_s = 5.640\,5$.

the correction of characteristics to unbiassed parameters is quite a difficult task, because the effect of an extreme on the extent of the random error, and thus also on the representativeness of the sample, can scarcely be estimated dependably in advance.

Characteristic examples of these relationships are shown in Figs 28 and 29, presenting the transgression lines of the sample coefficients of variation and asymmetry in 10 000-term random sequences, either with Pearson's distribution of the IIIrd type and less pronounced variance and skewness, or with log-normal distribution and more pronounced variance and skewness.*) The properties of the transgression lines of the characteristics of all the mathematical models examined are described in detail in Table 10.

The solution led to the following results:

1. The number of the statistically significant extremes increases with the parameters of distribution increasing, which is quite logical. For instance, with the input $C_v = 1.20–1.50$ and $C_s = 4C_v$, the number of the significant extremes exceeds the number of the extremes that are insignificant (Table 10).

2. The C_v and C_s transgression lines confirm a relatively considerable variance of the significant extremes, and variously numerous groups of the characteristics biassed by extremes also occur. The least numerous groups examined were three-element groups; with C_v they can virtually occur within the range of the whole transgression line, and with C_s approximately in its left part (i. e. up to the probability of transgression equal to approximately 50 percent). A certain tendency can be observed in the C_s transgression lines with higher C_s/C_v ratios, although here, too, the longest continuous group of samples with extremes at the significance level of 1 percent has only 153 members, i. e. approximately 30 percent of the whole number of samples (Table 10, the last but one row).

The properties of the other longest continuous groups of biassed characteristics were also of considerable interest. With the sample C_v's with extremes at significance level equal to 1 percent their continuous groups were less numerous, the relatively longest (with 52 sample C_v's) having been found in the series with log-normal distribution and output $C_v = 1.587\,8$ and $C_s = 6.080\,3$ (see Table 10). But neither in this most acceptable case can a rule be derived governing the verification of the C_v random errors biassed by extremes (52 cases equalling only 23 percent of 230 sample C_v's with extremes at significance level equal to 1 percent, and only 10 percent of the total number of 500 samples).

*) In view of the fact that the differences between the adjacent values of characteristics were insignificant, their lines of transgression were plotted from every fifth ordinate and the first largest, so that the transgression lines are shaped by 101 ordinates. Their representation of the alternation of the extremes of various significance is schematic.

TABLE 10. Properties of 30-element samples with extremes, assessed using the non-parametric Dixon's test in the lines of transgression

Type of distribution	Output parameters of the 10 000-element sequence		Number of samples: with extremes at signif. level 1 %	Number of samples: with extremes at signif. level 5 %	Number of samples: with insignificant extremes	Probability of transgression of the last three-element group of samples with extremes at the level of significance 1 %: with C_v	Probability of transgression …: with C_s	Number of elements of the longest continuous group with extremes at significance level 1 %: with C_v	Number of elements …: with C_s
Pearson's IIIrd type	$\bar{Q} = 1.013\,8$ $C_s = 1.593\,9$	$C_v = 0.760\,6$	100	61	339	52 %	38 %	4	12
Pearson's IIIrd type	$\bar{Q} = 1.050\,3$ $C_s = 4.491\,3$	$C_v = 1.248\,3$	239	54	207	94 %	58 %**)	14	74**)
Pearson's IIIrd type	$\bar{Q} = 1.038\,8$ $C_s = 4.334\,2$	$C_v = 1.229\,9$	230	70	200	62 %	55 %	13	94
Log-normal	$\bar{Q} = 1.012\,3$ $C_s = 0.775\,6$	$C_v = 0.746\,8$	51	30	419	4 %*)	16 %	7	6
Log-normal	$\bar{Q} = 1.068\,2$ $C_s = 6.080\,3$	$C_v = 1.587\,8$	230	34	236	51 %	46 %	52	153
Log-normal	$\bar{Q} = 1.056\,6$ $C_s = 5.640\,5$	$C_v = 1.548\,5$	2 218	55	227	77 %	53 %	16	107

*) a seven-element group,

**) found in a 293-element set of C_s

With sample C_s's the continuous groups were invariably longer, the longest with 153 elements, as already mentioned above.

The random occurrence of identified characteristics in nearly the whole domain of the transgression curve indicated that the effect of an extreme upon the magnitude of the random error could not be estimated reliably (i. e quantified) in advance, and that the estimation of the unbiassed parameters was therefore most difficult in these cases.

It is an interesting property of nearly all the samples with extremes at significance level equal to 1 percent that the standard variable of the extreme $t_{max} = (x_{max} - \bar{x})/s_x$ is greater than 3. This property can help to at least approximately estimate the significance of the extreme in actual cases.

From our examination of the problem it thus follows that in view of the occurrence of the extremes the estimation of unbiassed parameters proves to be rather difficult, and no general and reliable guidelines on the correction of the biassed characteristics can be formulated. As particularly doubtful must be viewed the estimation of the unbiassed coefficient of variation, where looser tendencies prevail in the formation of the groups of the sample coefficients of variation within the bounds of the lines of transgression. Account can only be taken of the bias of the sample coefficients of asymmetry with higher C_s/C_v ratios, which manifest certain tendencies to form groups of above-average values biassed by statistically significant extremes. In these cases it will be possible to reduce the theoretical systematic errors involved in the estimation of C_s.

We conclude the chapter by saying that the problems of parameter estimation with extremes occurring in the samples will mostly have to be tackled individually, and also that all the genetic aspects of the case under examination will need to be considered. Where the representativeness of the given sample biassed by an extreme is not assessable with acceptable certainty, the characteristics of that sample will have to be regarded as the expected values of the set of the characteristics and corrected to unbiassed parameters in accordance with current procedure and the principles of the theory of estimation.

5
Estimation of parameters by the method of maximum likelihood

5.1 Brief review of the development of the method

The sixty-plus years' development of the method of maximum likelihood, the idea of which was first conceived by R. Fisher, often raises hopes of achieving efficient and consistent estimators of the parameters of a universe from a single sample. The growing interest in the method is demonstrated by the large number of papers and articles published in journals, in which the process of estimation is studied from various aspects, particularly from the point of view of the type of probability distribution, the effect of the length of the sample on the estimation, and the errors of parameter estimation. Some of them take a fairly critical stance, pointing to the specific properties of the method, particularly to numerical problems and the limiting conditions of its application to actual cases.

Our interest therefore concentrated, besides the general mathematical treatises, upon the works available in the field of applied hydrology dealing with the problems of estimation of the paramaters of various types of flow series. As fundamental in this field must be considered the already quoted study by Matalas and Wallis [68] concerned with estimation of the parameters of the Pearson distribution of the IIIrd type. It analyses the limiting conditions of the validity of algorithms in estimation, and points to the pitfalls of the numerical solution, particularly as far as the shorter samples are concerned. It confirms the advantages of the method of maximum likelihood, which produces efficient estimates (estimates with minimum dispersion) as compared with the moments method. The authors also subject to critical analysis the causes of the limited spread of the method in hydrological practice, seeing the greatest obstacle to the application of the method in the complexity of the solution of the system of non-linear likelihood equations, in the difficulties of the numerical form of the estimation of parameters from shorter samples, and in the enhanced mathematical requirements of the method as compared with the requirements of the moments method. These circumstances are also considered to be the cause of the relatively small number of publicatons dealing with these problems. The authors voice their hope that the obstacles may be overcome with the help of modern computer technology and Monte Carlo methods of simulating random series.

In the West European literature on the subject, Condie's paper [27] is also most valuable. It derives the fundamental relationships for estimation of the parameters of the series with logarithmic Pearson's distribution of the IIIrd type, and in its application-oriented section it analyses 37 Canadian flow series of various lengths (24 to 64 years). The estimates of the *n*-year long maximum flows, with the help of the method of maximum likelihood and the comparison of these estimates with the results produced by the moments method, prove the method of maximum likelihood to be invariably less prone to random errors, and thus more suitable.

In the Soviet literature we found Fedorov's work published in 1960 [31] to be of considerable interest. Assuming Pearson's distribution of the IIIrd type and $C_s = 2C_v$, Fedorov computed the coefficients of variation of the annual flows in ten profiles of eight rivers in Kazakhstan. The values of C_v derived using the method of maximum likelihood are invariably a little lower than the values arrived at with the help of the moments method. Fedorov also mentions Kritskiĭ's and Menkel's work [59] from the year 1949, which may be regarded as the first postwar work of Soviet statistical literature dealing with the application of the method of maximum likelihood to the estimation of parameters of river run-offs.

In their latest book, published in 1981, Kritskiĭ and Menkel' [61] also give considerable attention to the method of maximum likelihood. After dealing with the general principle, they mention the qualities of the method. They consider it, compared with the moments method, as more advantageous, primarily because the variances of its estimates are markedly lower. And the work assumes that the estimators are practically unbiassed; it is therefore recommended that this quality should be checked. The authors deal in detail with the estimation of parameters for the triparametric gamma distribution, they quote the general algorithms to be used in estimation and mention the simplified procedures according to Blokhinov.

In 1968, a valuable and comprehensive study was published by Blokhinov [9]. Assuming the triparametric gamma distribution frequently used in the CIS and invariably referred to as Kritskiĭ–Menkel's, he derived the general expressions for the estimation of the parameters of this distribution by means of the method of maximum likelihood. Since the algorithms for parameter estimation are fairly complex, and a computer thus proves to be indispensable, but is not always readily available, the author also derived a simplified version, which involves estimating the parameters with the help of diagrams. Using these diagrams requires only the numerical expression of two of the values of the auxiliary statistical characteristics. The diagrams hold for values of the coefficient of variation within the following limits, $C_v = 0.25$ to 1.50 and ratio $C_s/C_v = 1$ to 6.

Another simplified procedure was derived by Blokhinov for the $C_s = 2C_v$ condition. In this case it is sufficient to estimate only the coefficient of variation from a single likelihood equation (the expected value of the sample means being unbiassed, the moments method should suffice to estimate it).

In 1970, Blokhinov in collaboration with Sotnikova published an interesting comparative study [10], in which the two authors evaluate the estimating performance of the method of maximum likelihood and the moments method as far as the coefficients of variation and asymmetry are concerned. The evaluation was based upon computations of C_v and C_s for 120 profiles of rivers in the European part of the CIS (length of observation: 40 years at least) and in Kazakhstan (25 to 30 years at least). Approximately up to the value of $C_v = 0.50$ the differences in the estimates produced by the two methods were extremely small; for higher values of C_v (in Kazakhstan) the estimates of the method of maximum likelihood (assuming $C_s = 2C_v$) were lower by 0.15 to 0.20. A similar result was arrived at by a comparison of C_v estimated using the method of maximum likelihood and C_s estimated individually, and C_v estimated with the help of the same method, but assuming $C_s = 2C_v$. It turned out that approximately up to the value of $C_v = 0.50$ the differences between the results of the two estimating procedures were negligible, and for the higher values of C_v (again in Kazakhstan) lower estimates were produced on the assumption of $C_s = 2C_v$.

An interesting result was obtained from a comparison of 120 values of ratio C_s/C_v estimated by both the method of maximum likelihood and the moments method. Prevailingly, lower values of that ratio were arrived at using the moments method, which can be accounted for by the bias of the moment estimates, particularly with the high values of C_v.

The estimation of C_s causes difficulties, particularly with shorter samples. The investigation of the correlation between C_s and C_v estimated with the help of the maximum likelihood method showed that with the majority of the profiles, ratio C_s/C_v occurred between values 2 and 3, particularly with the larger values of C_v. The mean value of the set of the values of ratio C_s/C_v found using the maximum likelihood method is 1.89 and 2.54 for the profiles in the European part of the CIS and Kazakhstan, respectively. The Soviet design practice often assumes that $C_s = 2C_v$, or it follows Kritskiĭ's and Menkel's recommendation to consider the so-called weighted value of C_s obtained from the computation of C_s for the given observation series and its correction, which is a function of the skewness estimated on the basis of the processed flow series of a number of rivers. This method of estimating C_s is thus characterized by a marked regional aspect.

The method of maximum likelihood is critically considered in Kartvelishvili's latest work [41]. The author formulates the principle of the method, as well as the advantages of its application to hydrological information. In some cases (e. g. of normal distribution) the method may yield estimates identical with those obtained using the moments method; in general, however, the computations of the maximum likelihood function may lead to a system of complex transcendental equations. Kartvelishvili therefore quotes Blokhinov [9] and draws the reader's attention to the possibility of simplified estimation making use of the diagrams for the triparametric gamma distribution. Aparrt from this, he also presents a simplified estimation procedure using the diagrams derived by G. A. Grinevich, Petelina and A. G. Grinevich [34].

The principles of the method, and its fundamental properties, have also been repeatedly dealt with in Czechoslovak statistical literature. For instance, Anděl [3] concentrates on the solution of the likelihood equations and the asymptotic distribution of the maximum likelihood estimates. Šor [110] shows the application of the method to a simple example of normal and truncated distributions; Kubáček [64] gives a general treatise on the consistency of the estimators.

In the Czechoslovak water-engineering literature the method was most extensively dealt with by Kos [54], who presented a simplified version of the method applied to the estimation of the coefficient of variation of the average annual flows, and who pointed to the method's practical importance as far as processing of hydrological information for the design of storage reservoirs is concerned.

The method has not been fully adopted in Czechoslovak water-engineering practice so far, although the Czechoslovak National Standard No. 73 6805 recommends using it for parameter estimation with larger flow series variances [28].

The evaluation of the development of the method of maximum likelihood exposes the urgency of further research in this field. What the Czechoslovak statistical literature lacks is a comprehensive work on the application of algorithms to the estimation of the parameters of hydrological series with the most frequent types of distribution. Also, the problems of the estimation of parameters from various random samples of the same population, the bias of the estimators, and the conditions of the numerical solution of the system of non-linear likelihood equations remain unsatisfactorily elucidated.

The method of maximum likelihood is a typical method, the adaptation of which to concrete cases is conditional upon a number of aspects being taken into consideration. Statistical literature, for instance, emphasizes the efficiency of the method as its advantage. The method is however considerably biassed – in the same way as the moments method. Our research was therefore also oriented towards these complex problems.

The chief aim of the research was to supply material for comparative evaluation of various methods of estimating the parameters of hydrological series, and for the improvement of the method of determining the hydrological design quantities desired.

5.2 Principle of the method of maximum likelihood and the application of simulation models of random sequences to estimation

The method of maximum likelihood is a general method of estimating the parameters of a universe from the given sample $x_1, x_2, \ldots, x_n$. Like the moments method, the method of maximum likelihood yields point estimates, but it differs in principle from the moments method quite substantially.

In theory and in practice, the method is primarily applied owing to two basic advantages: the method makes use of efficient estimators (i. e. estimators producing minimum variance); and the method's estimators are consistent (i. e. with the lengths of the samples increasing, the expected values of the parameters estimated from the set of samples converge towards the parameters of the universe).

The unknown parameters of the universe are derived under the assumption that only the analytical form of the equation of probability density $\varphi(x)$ of random variable x is known. This function contains the parameters sought (e. g. the mean, the coefficient of variation and the coefficient of asymmetry), so that function $\varphi(x)$ can be rewritten in the following form,

$$\varphi(x) = \varphi(x, s_1, \ldots, s_m),$$

where s_i are statistical parameters, $i = 1, 2, \ldots, m$ (number of parameters).

The probability of securing the given sample with the given density within the limits

for first term	from x_1 to $x_1 + \Delta x_1$,
for second term	from x_2 to $x_2 + \Delta x_2$
⋮	
for n-th term	from x_n to $x_n + \Delta x_n$,

is evidently equal to

$$p = \varphi(x_1)\,\varphi(x_2) \ldots \varphi(x_n)\;\Delta x_1\,\Delta x_2 \ldots \Delta x_n\,. \tag{5.1}$$

In functions $\varphi(x_i)$ the right-hand side of equation (5.1) contains the unknown parameters.

The most probable parameters are obviously those for which probability p is maximal. Finding the maximum of p for the product of the functions in (5.1) is facilitated if the logarithm of equation (5.1) is taken and the resulting equation rewritten in the following form:

$$L = \ln \frac{p}{\Delta x_1 \, \Delta x_2 \, \dots \, \Delta x_n} = \sum_1^n \ln \varphi(x_i) \,. \tag{5.2}$$

Function L is the likelihood function. It is obvious that for all the values of the parameters for which p is maximal, function L will also be maximal.

From this property equations can be derived for the estimation of the unknown parameters s_i. The extreme of function (5.2) is obtained from zero partial derivatives

$$\frac{\partial L}{\partial s_i} = 0 \,, \qquad i = 1, 2, \dots, m \,. \tag{5.3}$$

Formula (5.3) is referred to as the likelihood equation.*)

Solving this system will lead us to the required parameters s_i.

The principle of the method of maximum likelihood discussed above also points to the method of estimation from a single sample only. The right side of equation (5.2) is expressed first, then follows the derivation of the system of likelihood equations (5.3), and the numerical solution of the latter will give us parameters s_i. The problems involved in this solution will be dealt with below with respect to the individual types of $\varphi(x)$.

A problem of considerably greater complexity concerns the properties of the estimates that have been arrived at in this way on the basis of various samples of the same universe. It can easily be demonstrated that various estimates of the parameters of the same universe can be obtained from various random samples. The estimates are thus of a haphazard character. In order that the properties of such a set of estimates may be defined, it is indispensable that they should be statistically processed (in a way similar to that used by the moments method). This solution, several times repeated for the individual parameters and various lengths of the samples, enables us to find such properties of the set of estimates as, for instance, their mean values, variances, distributions of probability, as well as efficiency, consistency etc.

It is at first sight evident that the statistical problem described is rather complex and that tackling it with the help of analytical methods for the various types of distribution of the population is impracticable. Modelling methods can again prove to be of some help. If random samples (of various lengths) are generated from a sufficiently long modelled random sequence, the method of

---

*) The substantial conditions of the solution of system (5.3), the existence of its solution, and the limit distribution of the most likely estimates are dealt with in detail by Anděl [3].

maximum likelihood can be used to estimate the parameters sought, which are then statistically processed and compared with the output parameters of the random sequence representing the population.

In our research, we proceeded such that for the types of distribution selected we modelled 10 000-term random sequences, from which we formed 500 random samples of various length. We then estimated the parameters from each sample using the maximum likelihood method. The probability properties of the parameters estimated could thus be derived from 500-element sets, which were then processed with the help of the currently used moments method in the same way as with the random variables. For these sets of parameters we ascertained all the basic statistical characteristics, including the maximum and minimum elements of the sets as well as their critical values with probabilities of transgression equal to 97.5 percent and 2.5 percent expressing the confidence interval of the usual 5 percent level of significance.

This detailed procedure made it possible to satisfactorily assess the behaviour of the parameters estimated, above all their relationship to the parameters of the whole population (viz. the magnitude of the systematic errors and the consistency of the estimators).

At the same time, we compared the properties of the parameters estimated from the individual samples using the method of maximum likelihood with the results of the moments method. Both these methods were thus used to estimate the parameters for all the samples of random sequences, and the results were further statistically processed, as already mentioned above.

5.3 Estimation of parameters of populations with various probability distributions

5.3.1 Estimation of parameters of a population with Pearson's distribution of the IIIrd type

Our point of departure was the equation of Pearson's density of the IIIrd type in the following form [11]:

$$\varphi(x) = \frac{\alpha}{\Gamma(\lambda)} \left[\alpha(x - m)\right]^{\lambda - 1} e^{-\alpha(x-m)}, \tag{5.4}$$

where α, λ, m stand for the unknown parameters of density, explained in Fig. 30. Symbol Γ denotes function gamma; its argument must be positive, viz. $\lambda > 0$.

If the logarithm is taken of the two sides of equation (5.4), we get

$$\ln \varphi(x) = \ln \left(\frac{\alpha}{\Gamma(\lambda)}\right) + (\lambda - 1) \ln \alpha + (\lambda - 1) \ln (x_i - m) - \alpha(x_i - m). \tag{5.5}$$

From equation (5.5) follow the first limiting conditions of the maximum likelihood method with this type of distribution. Since $\Gamma(\lambda) > 0$, the condition of inequality $\alpha > 0$ (i. e. skewness must be positive) must simultaneously be satisfied, and $x_i - m > 0$ must hold for all i's.

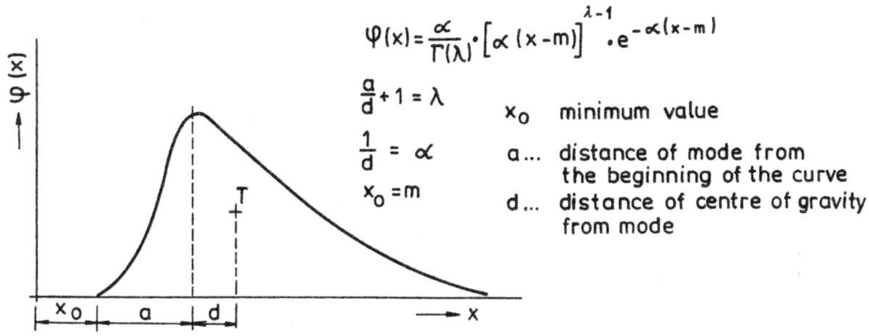

Fig. 30. Pearson's IIIrd type density – meaning of parameters.

The likelihood function (5.2) is derived in accordance with its definition as a sum of logarithms of all the terms of the given sample, in the following form:

$$L = n \ln \left(\frac{\alpha}{\Gamma(\lambda)}\right) + n(\lambda - 1) \ln \alpha + (\lambda - 1) \sum_1^n \ln (x_i - m) - \alpha \sum_1^n (x_i - m), \tag{5.6}$$

where n stands for the number of the terms of the sample.

The likelihood equation (5.3) is derived as partial derivatives of L with respect to parameters α, λ, m, which are put equal to zero.

$$\frac{\partial L}{\partial \alpha} = n \frac{\partial}{\partial \alpha} [\ln \alpha - \ln \Gamma(\lambda)] + \frac{n(\lambda - 1)}{\alpha} - \sum_1^n (x_i - m) =$$

$$= \frac{n}{\alpha} + \frac{n(\lambda - 1)}{\alpha} - \sum_1^n (x_i - m),$$

$$\frac{n\lambda}{\alpha} - \sum_1^n (x_i - m) = 0. \tag{5.7}$$

$$\frac{\partial L}{\partial \lambda} = -n\frac{\partial}{\partial \lambda}\left[\ln \Gamma(\lambda)\right] + n \ln \alpha + \sum_1^n \ln (x_i - m),$$

$$-n\frac{\partial}{\partial \lambda}\left[\ln \Gamma(\lambda)\right] + n \ln \alpha + \sum_1^n \ln (x_i - m) = 0. \tag{5.8}$$

$$\frac{\partial L}{\partial m} = (\lambda - 1)\frac{\partial}{\partial m}\sum_1^n \ln (x_i - m) - \alpha\frac{\partial}{\partial m}\sum_1^n (x_i - m) =$$

$$= (\lambda - 1)\sum_1^n \frac{-1}{x_i - m} - \alpha(-n),$$

$$(\lambda - 1)\sum_1^n \frac{-1}{x_i - m} + \alpha n = 0. \tag{5.9}$$

The non-linear character of the likelihood equations (5.7), (5.8), (5.9) indicates that the computation of the unknown parameters α, λ, m will be rather difficult and will require adequate numerical (iterative) methods and an efficient computer. The need to repeat this procedure for each random sample poses a difficult problem of application, the solution of which has so far evidently been the main obstacle to the maximum likelihood method being more extensively used in practice, and to the qualities of that method being fully elucidated.

Our research showed that the method of steepest descent with numerical estimation of the magnitude of the derivatives was the most applicable to the solution of the given system of non-linear equations.

The statistical parameters $\bar{X}$, C_v, C_s of the population can be derived from density parameters α, λ, m according to the following relationships [11, 77, 80]:

expected value:

$$\bar{X} = m + \frac{\lambda}{\alpha}, \tag{5.10}$$

coefficient of variation:

$$C_v = \frac{\alpha}{|\alpha|}\frac{\lambda^{1/2}}{(\lambda + m\alpha)}, \tag{5.11}$$

coefficient of asymmetry:

$$C_s = \frac{\alpha}{|\alpha|}\frac{2}{\lambda^{1/2}}. \tag{5.12}$$

Since equation (5.9) holds for all the samples, and since inequalities $\alpha > 0$, $(x_i - m) > 0$ must be satisfied, condition $\lambda > 1$ must simultaneously also be fulfilled. From equation (5.12) we thus get another limiting condition of the maximum likelihood method for the given distribution, viz. $C_s < 2$, which holds for all the samples. And the estimate of C_s of the population with the help of the expected value of the set of estimates of C_s from the samples must thus not exceed the value of $C_s = 2$, either. The method will therefore produce biassed estimates, with the systematic errors growing proportionally to skewness, as with the moments method.

From equation (5.11) it follows that the estimation of C_v will also be affected by the biassed estimators of C_s.

On the other hand, it can easily be shown from equation (5.10) with the help of the relationships presented in Fig. 30*) that the estimate of the expected value $\bar{X}$ coincides with the moments estimate. A satisfactorily large set of sample mean values can thus give us an unbiassed estimator of the mean value of the population.

The algorithms derived indicate the scope of application of the method of maximum likelihood with the given distribution to hydrological practice. Since the estimate of C_s must not exceed the value of 2, the method will be more applicable to the estimation of the parameters of the annual hydrological series, not to the estimation of the parameters of the culminating flow series, which invariably exhibit higher fluctuation and skewness.

Researching and assessing the behaviour of the parameters estimated from various random samples with the help of the maximum likelihood method, we made use of the usual procedures of the moments method. For the results the reader is referred to Section 5.4.

5.3.2 Estimation of parameters of a population with logarithmic Pearson distribution of the IIIrd type

The likelihood function and the likelihood equations can be derived similarly as with the Pearson distribution of the IIIrd type. The logarithmic Pearson density of the IIIrd type can be expressed in the following form:

$$g(x) = \frac{\alpha}{\Gamma(\lambda)}[\alpha \ln x - m)]^{\lambda-1}\, e^{-\alpha(\ln x - m)}\, \frac{1}{x}, \tag{5.13}$$

where α, λ, m again stand for the unknown parameters. Equation (5.13) follows immediately from equation (4.18) for $k = 1$.

---

*) Simple modification will change the form of equation (5.10) to $\bar{X} = m + a + d$.

The natural logarithm of function (5.13) is

$$\ln g(x) = \ln\left(\frac{\alpha}{\Gamma(\lambda)}\right) + (\lambda - 1)\ln\alpha +$$

$$+ (\lambda - 1)\ln(\ln x - m) - \alpha(\ln x - m) - \ln x\,. \tag{5.14}$$

The likelihood function equals the sum of the logarithms of all the terms of the given sample, viz.

$$L = n\ln\left(\frac{\alpha}{\Gamma(\lambda)}\right) + n(\lambda - 1)\ln\alpha + (\lambda - 1)\sum_1^n \ln(\ln x_i - m) -$$

$$- \alpha\sum_1^n (\ln x_i - m) - \sum_1^n \ln x_i\,, \tag{5.15}$$

where n denotes the number of the terms of the sample.

The likelihood equations can be derived as follows:

$$\frac{\partial L}{\partial\alpha} = n\frac{\partial}{\partial\alpha}[\ln\alpha - \ln\Gamma(\lambda)] + \frac{n(\lambda - 1)}{\alpha} - \sum_1^n (\ln x_i - m) =$$

$$= \frac{n}{\alpha} + \frac{n(\lambda - 1)}{\alpha} - \sum_1^n (\ln x_i - m)\,,$$

$$\frac{n\lambda}{\alpha} - \sum_1^n (\ln x_i - m) = 0\,. \tag{5.16}$$

$$\frac{\partial L}{\partial\lambda} = -n\frac{\partial}{\partial\lambda}[\ln\Gamma(\lambda)] + n\ln\alpha + \sum_1^n \ln(\ln x_i - m)\,,$$

$$-\frac{\partial}{\partial\lambda}[\ln\Gamma(\lambda)] + n\ln\alpha + \sum_1^n \ln(\ln x_i - m) = 0\,. \tag{5.17}$$

$$\frac{\partial L}{\partial m} = (\lambda - 1)\frac{\partial}{\partial m}\left[\sum_1^n \ln(\ln x_i - m)\right] - \alpha\frac{\partial}{\partial m}\left[\sum_1^n(\ln x_i - m)\right] =$$

$$= (\lambda - 1)\sum_1^n \frac{-1}{\ln x_i - m} - \alpha(-n)\,,$$

$$(\lambda - 1)\sum_1^n \frac{-1}{\ln x_i - m} + \alpha n = 0\,. \tag{5.18}$$

Equations (5.16), (5.17) and (5.18) are formally analogous with equations (5.7), (5.8) and (5.9). They differ only by the logarithms of the given values of x_i, which is quite logical considering the relationship between the Pearson and the logarithmic Pearson distribution.

From the character of equations (5.16), (5.17), and (5.18) it is evident that the computation of the unknown parameters α, λ, m is more exacting than the analogous procedure with Pearson's distribution. And the task becomes even more difficult if the examination concentrates on the behaviour of the parameters estimated from a set of random samples of the same population.

The forms of the limiting conditions of the validity of the equations derived are also reminiscent of those pertaining to the Pearson distribution. From the likelihood function (5.15) it can above all be seen that inequalities $\alpha > 0$ and $\lambda > 0$ must be satisfied. For all the terms of the samples it must hold that $x_i > 0$ (the whole density lying in quadrant 1). The most radical limitation follows from the inequality $(\ln x_i - m) > 0$, which must also be satisfied for all i's. From this condition, and from equations 5.12 and 5.18, relationships $\ln x_i > m \Rightarrow \lambda > 1 \Rightarrow C_s < 2$.

These limiting conditions at the same time imply limited possibilities of applying the method to the concrete tasks of hydrological practise, which are analogous to those pertaining to the Pearson distribution.

From the parameters of density α, λ, m estimated the statistical parameters of the logarithms of variables x_i (i. e. parameters of Pearson's distribution) are simply derived according to equations (5.10), (5.11) and (5.12). The derivation of the parameters of the given variables x_i (i. e. parameters of logarithmic Pearson's distribution) is much more complex. Bobée [11] expresses the relationship between the first three general moments and the parameters of density α, λ, m by a system of equations (4.24), (4.25) and (4.26), which is then handled in the way described in Chapter 4.

5.3.3 Estimation of parameters of a population with normal and log-normal distributions

In solving this problem we first tackled the estimation of the statistical parameters of Gaussian distribution. It can easily be shown (e. g. [110]) that in this case estimation with the help of the method of maximum likelihood is coincident with estimation by means of the moments method.

Let the density of Gaussian distribution be given in the following form:

$$\varphi(x) = \frac{1}{\sigma\sqrt{2\pi}} \exp\left[-\frac{(x-\mu)^2}{2\sigma^2}\right]. \tag{5.19}$$

Then its natural logarithm is equal to

$$\ln \varphi(x) = -\ln \sigma - \ln \sqrt{2\pi} - \frac{1}{2\sigma^2}(x_i - \mu)^2 , \tag{5.20}$$

and the likelihood function can be expressed in the following form:

$$L = -n \ln \sigma - n \ln \sqrt{2\pi} - \frac{1}{2\sigma^2} \sum_1^n (x_i - \mu^2) . \tag{5.21}$$

For the two unknown parameters μ and σ the likelihood equations are derived in the usual way. They then give the following well-known expressions for μ and σ:

$$\mu = \frac{1}{n} \sum_1^n x_i , \tag{5.22}$$

$$\sigma^2 = \frac{1}{n} \sum_1^n (x_i - \mu)^2 , \tag{5.23}$$

which coincide with the definitions for the first general and second central moment.

The parameters of the log-normal distribution have analogous properties. Let the density of that distribution be given in the following form:

$$\varphi(x) = \frac{1}{(x - x_0)\, \sigma(y) \sqrt{2\pi}} \exp \left\{ \frac{-[\ln (x - x_0) - \mu(y)]^2}{2\sigma^2(y)} \right\} , \tag{5.24}$$

where x_0 denotes the minimum term of the series. Its natural logarithm is given by:

$$\ln \varphi(x) = -\ln (x - x_0) - \ln \sigma(y) - \ln \sqrt{2\pi} - \\ - \frac{1}{2\sigma^2(y)} [\ln (x - x_0) - \mu(y)]^2 , \tag{5.25}$$

and the likelihood function can be expressed as follows:

$$L = - \sum_1^n \ln (x_i - x_0) - n \ln \sigma(y) - n \ln \sqrt{2\pi} - \sum_1^n \frac{[\ln (x_i - x_0) - \mu(y)]^2}{2\sigma^2(y)} . \tag{5.26}$$

For the three unknown parameters $\mu(y)$, $\sigma(y)$, x_0 the following expressions can be derived from the likelihood equations:

$$\mu(y) = \frac{1}{n} \sum_1^n \ln (x_i - x_0), \tag{5.27}$$

$$\sigma^2(y) = \frac{1}{n} \sum_1^n [\ln (x_i - x_0) - \mu(y)]^2 . \tag{5.28}$$

The minimum terms, x_0, in the population must be estimated by iteration from the likelihood equation

$$0 = \sum_1^n \frac{1}{x_i - x_0} \left[1 - \frac{\mu(y)}{\sigma^2(y)} \right] + \frac{1}{\sigma^2(y)} \sum_1^n \frac{\ln (x_i - x_0)}{x_i - x_0} . \tag{5.29}$$

Since equations (5.27) and (5.28) are, like equations (5.22) and (5.23), moment estimators, it will suffice to concentrate upon the examination of the properties of the minimum term, x_0, estimated from equation (5.29). As in the preceding task, it is advisable that use should be made of the random samples of modelled sequences, with the results statistically processed.

5.3.4 Estimation of parameters of a population with triparametric gamma distribution

This task poses the most difficult problems, for the analytic form of the density of the triparametric gamma distribution is the most complex. Blokhinov [9] bases his method of solution upon the expression of the denstity in the following form:

$$p(x; x'; \gamma; b) =$$
$$= \begin{cases} \dfrac{\Gamma(\gamma + b)^{\gamma/b}}{\Gamma(\gamma)} \dfrac{1}{|b|\, \Gamma(\gamma)\, x'^{\gamma/b}} x^{(\gamma/b)-1} \exp\left[-\left(\dfrac{\Gamma(\gamma + b)}{\Gamma(\gamma)} \dfrac{x}{x'} \right)^{1/b} \right] & \text{for} \quad x > 0 \\ 0 & \text{for} \quad x = 0 . \end{cases} \tag{5.30}$$

where x' stands for the expected values, and γ, b for the other parameters of the distribution linked with the coefficient of variation C_v and the coefficient of asymmetry C_s by transcendental relationships, viz.

$$C_v = \left[\frac{\Gamma(\gamma)\,\Gamma(\gamma + 2b)}{\Gamma^2(\gamma + b)} - 1\right]^{1/2}, \tag{5.31}$$

$$C_s = \frac{\dfrac{\Gamma^2(\gamma)\,\Gamma(\gamma + 3b)}{\Gamma^3(\gamma + b)} - 3\dfrac{\Gamma(\gamma)\,\Gamma(\gamma + 2b)}{\Gamma^2(\gamma + b)} + 2}{\left[\dfrac{\Gamma(\gamma)\,\Gamma(\gamma + 2b)}{\Gamma^2(\gamma + b)} - 1\right]^{3/2}}, \tag{5.32}$$

where $\Gamma(\gamma)$, $\Gamma(\gamma + b)$ and the other functions are gamma functions with the corresponding arguments.

Blokhinov makes use of the current procedure to derive the likelihood function L and from it the corresponding likelihood equations for the three unknown parameters x', γ and b, which he rewrites after some modification as

$$\left(\frac{\Gamma(\gamma + b)}{\Gamma(\gamma)}\right)^{1/b} \lambda_1 - \gamma = 0, \tag{5.33}$$

$$\lambda_2 + \ln\frac{\Gamma(\gamma + b)}{\Gamma(\gamma)} - b\frac{\partial}{\partial\gamma}\ln\Gamma(\gamma) = 0, \tag{5.34}$$

$$\gamma\lambda_2 - \left(\frac{\Gamma(\gamma + b)}{\Gamma(\gamma)}\right)^{1/b} \lambda_3 + b = 0, \tag{5.35}$$

where the following denotations apply:

$$\lambda_1 = \frac{\sum_1^n\left(\dfrac{x_i}{x'}\right)^{1/b}}{n}, \qquad \lambda_2 = \frac{\sum_1^n \ln\dfrac{x_i}{x'}}{n}, \qquad \lambda_3 = \frac{\sum_1^n\left(\dfrac{x_i}{x'}\right)^{1/b}\ln\dfrac{x_i}{x'}}{n}. \tag{5.36}$$

If we compare equations (5.33), (5.34) and (5.35) with the respective equations for the Pearson, or logarithmic Pearson, distribution, it turns out that the estimation of the parameters of gamma distribution x', γ and b is an extraordinarily difficult task, the solution of which requires special procedures and a powerful computer. But such a computer may not ensure full success in this matter, for the solution depends upon a number of circumstances, e. g. the different probability properties of the random samples and their lengths, limiting the validity conditions of the likelihood equations, numerical precision, and the rapidity of the convergence of the iterative procedures selected, the stability of the iterations etc.

In view of these difficulties, and also considering the fact that modern computer technology may not always be available, Blokhinov derived a simplified procedure of estimating parameters with the help of diagrams. The application of these diagrams requires only the numerical expression of the following two auxiliary statistical characteristics:

$$\lambda_2 = \frac{\sum_1^n \ln \frac{x_i}{x'}}{n}, \qquad \lambda_3' = \frac{\sum_1^n \frac{x_i}{\bar{x}} \ln \frac{x_i}{\bar{x}}}{n}, \tag{5.37}$$

and the value of C_v and C_s sought can then readily be read from the diagrams. As far as the expected value of x' is concerned, a satisfactory moment estimate is expected.

The possibilities of the diagrams being used for the estimation of C_v and C_s were assessed with several random sequences and their samples. It turned out that the diagrams were not derived for a sufficiently wide range of input characteristics λ_2 and λ_3', so that their application to our conditions was rather limited.

For the results of these investigations the reader is referred to the following Section, 5.4.

Blokhinov also studied other possibilities for simplifying the estimation of parameters related to an arbitrary ratio C_s/C_v. This, however, turned out to be no less difficult than the general solution. Only when $C_s = 2C_v$, does a single likelihood equation prove sufficient for estimating C_v, viz.

$$\frac{\partial}{\partial \gamma} \ln \Gamma(\gamma) - \ln \gamma - \lambda_2 = 0\,, \tag{5.38}$$

where $\gamma = 1/C_v^2$, λ_2 having the same meaning as in (5.37).

Blokhinov also mentions some works of statistical literature presenting graphical or tabular versions of the relationship of λ_2 and C_v facilitating practical computations. For hydrological practice, the assumption $C_s = 2C_v$ is of substantial importance, which need of course not always be satisfied by the series of variables.

5.4 Properties of parameter estimates of populations with various probability distributions

5.4.1 Properties of parameter estimates of a population with Pearson's distribution of the IIIrd type

It was the original task of our research to test the parameter estimators and the law-governed behaviour of the characteristics on a larger number of modelled sequences with various inputs. The limiting conditions of the method of maximum likelihood, particularly the $C_s < 2$ condition, as well as the numerical difficulties, however, necessitated reducing the number of variants to seven characteristic models with lower values of C_v and C_s. The input and output parameters are listed in Table 11.

TABLE 11. Input and output parameters of 10 000-element random sequences with Pearson's IIIrd type distribution selected for the examination of estimation using the method of maximum likelihood

Variant No.	Input parameters			Output parameters		
	$\bar{X}$	C_v	C_s	$\bar{X}$	C_v	C_s
1	1.00	0.31	1.00	1.000	0.324	1.033
2	1.00	0.50	0.50	1.006	0.506	0.511
3	1.00	0.75	0.75	1.001	0.765	0.832
4	1.00	0.50	1.50	1.000	0.514	1.541
5	1.00	0.75	1.50	1.008	0.765	1.588
6	1.00	1.00	1.00	1.005	1.012	0.962
7	1.00	1.20	1.20	0.995	1.247	1.200

The main outcomes of this study are as follows:

1. From the 500 random samples of 10 000-term modelled sequences a certain number always give unsatisfactory estimates of parameters; because the iterations are sensitive to the input parameters of the models, the dependence upon the original, i. e. starting, conditions is quite strong, for the estimates from some of the samples not even a relatively high number of iterations, 250 to 300, proves to be sufficient, some iterations converge towards unrealistic values, or do not converge at all, etc. The parameter estimates from a set of samples of various lengths are therefore hardly comparable, for the number of samples that each variant comprises differs considerably. These numerical difficulties of the method can particularly manifest their adverse effect wherever parameters are estimated from a single sample, which need not lead to a satisfactory solution.

2. With the method of maximum likelihood, estimating parameters from a single random sample can lead to rather uncertain results. Long-term para-

meters can only be approximated by the expected values of the estimates from the individual samples. The estimation of the means coincides with the estimation with the help of the moments method, and the estimators of the coefficients of variation and asymmetry can be biassed, they however invariably lead to

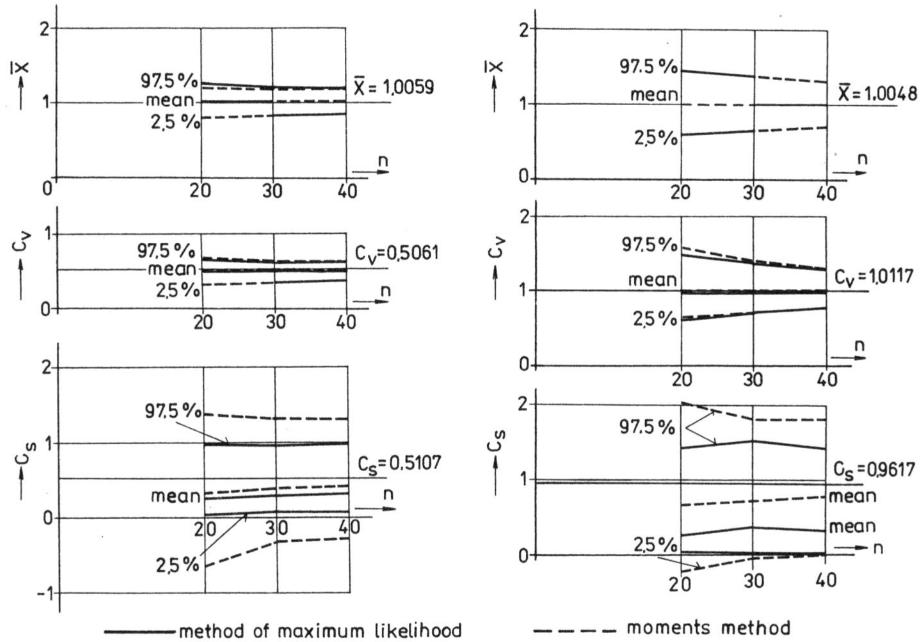

Fig. 31. Confidence intervals of sample means and sample coefficients of variation and asymmetry in a 10 000-element random sequence with Pearson's IIIrd type distribution obtained using the method of maximum likelihood and the moments method.

lower variance. This advantage of the method (viz. efficiency) can prove to be particularly effective in the case of the estimation of parameters from shorter samples with extreme terms (the estimates being lower than in the case of the moments method, in which parameters are sensitive to extreme numbers of samples).

A typical example of these relationships is presented in Fig. 31, which compares the results achieved by the two methods in the estimation of parameters of models No. 2 and 6 of Table 11. The estimates of the means are nearly coincident in the whole course of confidence intervals; the estimates of the coefficients of variation with the help of the method of maximum likelihood exhibit lower variance than the estimates produced using the moments method. The relatively most marked reduction of variance of the estimates of the coefficient of asymmetry is achieved by the method of maximum likelihood.

3. The estimates of parameters with the help of the maximum likelihood method are consistent within the bounds of the lengths selected, as in the case

of the moments method. The systematic errors of the parameter estimates ascertained with the help of the method of maximum likelihood are however in nearly all the cases examined, higher than in the case of the moments method (comp. Fig. 31, which visualizes this property on a curve of the expected values of C_s). The bias of the estimators proved can be regarded as quite a serious disadvantage of the method of maximum likelihood, because no methodological means or aids (as for example the diagrams, as in the moments method) are available for fast correction of the parameters. Without these aids, correction requires a complex and costly solution (with the help of mathematical models), which is hardly applicable to engineering practice without extraordinary measures being taken.

In the context of estimation bias we also studied the rather complex problem of the representation of the sample coefficients of asymmetry, which in the moments method can acquire positive as well as negative values, up to the interval (0;2) estimated by the method of maximum likelihood. With this representation we were not able to detect any regular tendencies or trends in how the values estimated by one method project themselves on the values estimated by the other method. The detection of such tendencies or trends would most probably require specific mathematical research.

With the given type of distribution the research virtually proved a single advantage of the method of maximum likelihood only, viz. lesser random errors and considerable estimating efficiency. This advantage is however outweighed by numerous disadvantages, particularly by the numerical precision and the estimating bias, which will most probably prevent the method from being more widely used in practice.

5.4.2 *Properties of parameter estimates of a population with logarithmic Pearson distribution of the IIIrd type*

Here, our point of departure were the equations (5.16), (5.17) and (5.18). The aim was again to estimate the parameters from a set of 500 random samples of the modelled series, and to define their behaviour.

The numerical difficulties of the maximum likelihood method had already manifested themselves with the Pearson distribution; they were vastly greater with the logarithmic Pearson distribution. Despite a number of attempts, we did not succeed in finding any numerical methods and procedures leading to acceptable and usable results. It is evident that the cause of that failure must be inherent in the essence of the method itself, in the form of the likelihood equations and in the different probability properties of the samples, which need not always satisfy the conditions limiting the solution.

Our research also covered an attempt at estimating the parameters of the logarithmic Pearson distribution with the help of a modelled sequence with Pearson's distribution and its samples. The computations were undertaken so that instead of the sequence with the logarithmic Pearson distribution, the sequence with Pearson's distribution was modelled first, and for the latter's samples parameters were estimated using both the moments method and the method of maximum likelihood. From the parameters of density α, λ, estimated, simultaneous estimates of the parameters of a sequence with the logarithmic Pearson distribution were made for each sample according to equations (4.24), (4.25) and (4.26) using the moments method as well as the method of maximum likelihood. Whereas the likelihood estimators for the Pearson distribution were successful, for the logarithmic Pearson distribution they were again numerically unstable and gave quite unreal results. The experiment confirmed the disadvantages of the method of maximum likelihood, and the experience gained showed that for the given type of distribution estimation with the help of the moments method – with the respective correction – was much more feasible.

The application of the logarithmic Pearson distribution to the processing of hydrological series has its advantages, which have been mentioned in chapter 4. Judging from our experience with the method of maximum likelihood, we agree with the recomendation by the American Water Resources Council of 1967 that this type of distribution be used for analysing the frequency of the occurrence of floods, linked with the method of moments [11, 27].

5.4.3 Properties of parameter estimates of a population with a log-normal distribution

For the reasons mentioned in paragraph 5.3.3, we concentrated on the examination of the properties of the minimum term x_0 in the transformational relation $y = \ln(x - x_0)$.

Three random 10 000-term sequences were modelled with the following inputs:

$$\bar{X} = 1\,, \quad C_v = 0.75\,, \quad C_s = 1.50\,, \quad x_0 = -0.608\,,$$

$$\bar{X} = 1\,, \quad C_v = 1.20\,, \quad C_s = 4.80\,, \quad x_0 = -0.068\,,$$

$$\bar{X} = 1\,, \quad C_v = 0.50\,, \quad C_s = 2.00\,, \quad x_0 = 0.159\,.$$

From each sequence 500 random samples were formed of a length of 20 terms (viz. years) each. For each sample the minimum terms x_0 were estimated using the maximum likelihood method as well as the moments method, these terms having been further compared with the "empirical" minimum terms ascertained in the random samples. These three sets of 500 x_0 values were then plotted to give the lines of transgression shown in Fig. 32. For all the samples the moment characteristics were also computed.

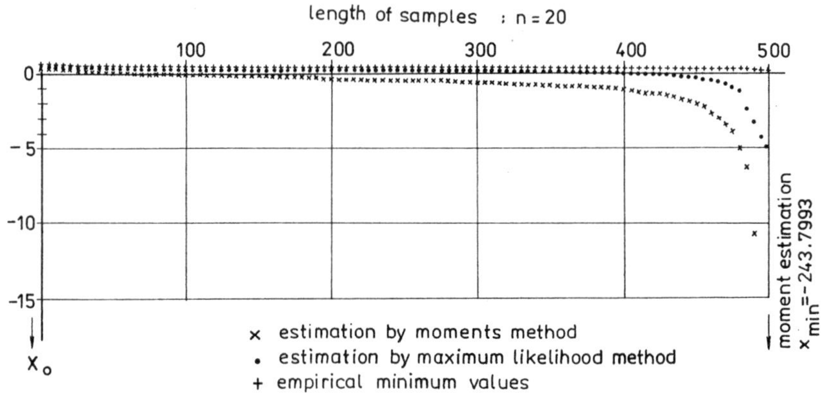

Fig. 32. Lines of transgression of the values of minimum terms x_0.
Inputs of model: $\bar{X} = 1.00$, $C_v = 0.50$, $C_s = 2.00$, $x_0 = 0.159$.

Three conclusions were formed from this study:

1. The estimates of the x_0 terms using the method of maximum likelihood do not markedly differ within the bounds of the lines of transgression up to a relatively high probability limit (the lines of transgression having a constant course in that section), and they do not differ markedly from the minimum term of the whole random series either. A marked decline in the x_0 values can be seen with samples of low skewness, the distribution of which approximates to Gaussian distribution. This result of the examination is quite logical and it corresponds to the well-known fact that log-normal distribution is suitable for sets with higher skewness only.

2. The estimates of the values of the x_0 terms with the help of the moments method are lower throughout the whole length of the lines of transgression than the estimates produced using the maximum likelihood method, and in considerable parts of these lines they differ markedly from the long-term x_0 value. In view of this fact, the method of maximum likelihood appears to be the more suitable method of estimating x_0 terms.

3. The investigation of the course of the lines of transgression of the minimum "empirical" terms did not yield any new information.

The problem of estimating the values of the minimum terms in the real series of hydrological variables can be quite complex unless the distribution of these variables is at least approximately log-normal. No general directions exist for estimating the terms in these cases; in practice, the minimum terms are therefore invariably extrapolated by approximative methods (tentatively, graphically in probability networks). The problem should receive full attention, because the solution of a number of important tasks, including mathematical modelling of flow series (see Chapter 11), is fully conditional upon the results in this line.

5.4.4 Properties of parameter estimates of a population with a triparametric gamma distribution

Since a general solution of the problem of estimating all the three parameters of this type of distribution is an extraordinarily complex numerical undertaking, we decided to make use of Blokhinov's diagrams. The aim was to generate 500 random samples of several modelled sequences, for which the coefficients of variation C_v and coefficients of asymmetry C_s could be estimated using the diagrams mentioned. It was our intention to subject again the probability properties of the parameters estimated in this way to statistical analysis aided by a computer, particularly as far as their measure of bias is concerned..

Using Klibashev and Goroshkov's tables [50] we modelled two random 10 000-term sequences for triparametric gamma distribution, with the following inputs:

$\bar{X} = 1\,,\ C_v = 1.00\,,\ C_s = 3.00\,,$ with the length of the samples equal to 20 and 40 terms (years);

$\bar{X} = 1\,,\ C_v = 1.50\,,\ C_s = 6.00\,,$ with the length of the samples equal to 40 terms (years).

For all the random samples the fundamental moment characteristics and the auxiliary statistical characteristics (5.37) were calculated with the help of a computer.

Using Blokhinov's diagrams, we found that their applicability was limited, for they did not cover the required range of the values of auxiliary characteristics λ_2 and λ_3', so that from the set of 500 samples only a small part of parameters C_v and C_s were estimable. These incomplete sets were not further processed statistically.

Characteristics λ_2 and λ_3' exhibit interesting probability properties, particularly the range of variation and the expected values, which were ascertained for three selected model variants in the sets of 500 samples. The results are given in Table 12.

Blokhinov's diagrams cover the variation of the λ_2 values within the limits $(-0.10;\ -1.00)$ and up to the value of $\lambda_3' = 0.40$. Comparing the diagrams with the results listed in Table 12, we can see that the range of the diagrams would

TABLE 12. Properties of auxiliary characteristics λ_2 and λ'_3 in 500-element sets of random samples

Input of models	Length of samples	λ_2			λ'_3		
		max. value	min. value	mean value	max. value	min. value	mean value
$\bar{X} = 1,\ C_v = 1.00,\ C_s = 3.00$	20	−0.128	−0.865	−0.383	0.854	0.107	0.345
$\bar{X} = 1,\ C_v = 1.00,\ C_s = 3.00$	40	−0.161	−0.757	−0.397	0.752	0.166	0.363
$\bar{X} = 1,\ C_v = 1.50,\ C_s = 6.00$	40	−0.321	−1.397	−0.664	1.383	0.273	0.597

have to be extended in order that parameters may be estimated within a wider interval, in which they also most often occur. The exacting operation of extending the applicability of the diagrams was however already outside the scope of our planned research.

The results of the investigation of parameter estimation of triparametric gamma distribution need in no way lead to any scepticism, if we consider that for this type of distribution parameters can, if required, easily be estimated using the moments method and corrected according to Rozhdestvenkiĭ's diagrams.

6
Estimation of parameters by the quantiles method

6.1 Principle of the quantiles method and the application of simulation models of synthetic sequences to estimation

The principle of the quantiles method has been sufficiently described in the water-engineering literature (e. g. [1, 30, 116, 117]) for various types of probability distribution. We are therefore going to mention briefly only the procedure devised by Alekseev.

In the quantiles method, estimation is based upon several pre-selected empirical quantiles x_i, which are read from the empirical curve of transgression for the given probabilities P_i. Other basic data are the standardized quantiles of the respective transgression curves. The unknown values of parameters are sought (estimated) in view of the condition that the theoretical curve of transgression defined by these values should cross the empirical quantiles selected. This methodological procedure thus contributes to the theoretical curve fitting closely the empirical line of the given sample, thus also to the detection of a fitting sample distribution. But a substantial question still remains unanswered, viz. whether the parameters derived in this way have the weight of unbiassed parameters of the population with different types of distribution.

With Pearson's type III distribution, three probabilities of transgression are selected, viz. $P_1 = 5\,\%$, $P_2 = 50\,\%$, and $P_3 = 95\,\%$, and the corresponding empirical quantiles x_1, x_2 and x_3 ascertained.

The auxiliary variable (the so-called index of skewness)

$$S = \frac{x_1 + x_3 - 2x_2}{x_1 - x_3} = f(C_s)\,, \tag{6.1}$$

is then computed and the corresponding value of the coefficient of asymmetry, C_s, read from Table 13, which also applies to the values of theoretical quantiles $\phi_5 - \phi_{95} = \phi_1 - \phi_3$ and $\phi_{50} = \phi_2$.

TABLE 13. Theoretical standardized quantiles of the Pearson distribution

C_s	$\Phi(P, C_2)$		S	C_s	$\Phi(P, C_s)$		S
	Φ_5–Φ_{95}	Φ_{50}			Φ_5–Φ_{95}	Φ_{50}	
0.0	3.28	0.00	0.00	2.7	2.74	−0.38	0.74
0.1	3.28	−0.02	0.03	2.8	2.71	−0.39	0.76
0.2	3.28	−0.03	0.06	2.9	2.68	−0.39	0.78
0.3	3.27	−0.05	0.08	3.0	2.64	−0.40	0.80
0.4	3.27	−0.07	0.11	3.1	2.62	−0.40	0.81
0.5	3.26	−0.08	0.14	3.2	2.59	−0.41	0.83
0.6	3.25	−0.10	0.17	3.3	2.56	−0.41	0.85
0.7	3.24	−0.12	0.20	3.4	2.53	−0.41	0.86
0.8	3.22	−0.13	0.22	3.5	2.50	−0.41	0.87
0.9	3.21	−0.15	0.25	3.6	2.48	−0.42	0.89
1.0	3.20	−0.16	0.28	3.7	2.45	−0.42	0.90
1.1	3.17	−0.18	0.31	3.8	2.43	−0.42	0.91
1.2	3.16	−0.19	0.34	3.9	2.41	−0.41	0.92
1.3	3.14	−0.21	0.37	4.0	2.40	−0.41	0.92
1.4	3.12	−0.22	0.39	4.1	2.38	−0.41	0.93
1.5	3.09	−0.24	0.42	4.2	2.36	−0.41	0.94
1.6	3.07	−0.25	0.45	4.3	2.34	−0.40	0.94
1.7	3.04	−0.27	0.48	4.4	2.32	−0.40	0.95
1.8	3.01	−0.28	0.51	4.5	2.30	−0.40	0.96
1.9	2.98	−0.29	0.54	4.6	2.28	−0.40	0.97
2.0	2.95	−0.31	0.57	4.7	2.26	−0.40	0.97
2.1	2.92	−0.32	0.59	4.8	2.23	−0.39	0.98
2.2	2.89	−0.33	0.62	4.9	2.21	−0.39	0.98
2.3	2.86	−0.34	0.64	5.0	2.18	−0.38	0.98
2.4	2.82	−0.35	0.67	5.1	2.15	−0.38	0.98
2.5	2.79	−0.36	0.69	5.2	2.15	−0.37	0.98
2.6	2.76	−0.37	0.72				

The unknown parameters of the distribution are computed with the help of the following expressions:

$$\bar{X} = x_2 - \sigma_x \phi_2 , \tag{6.2}$$

$$\sigma_x = \frac{x_1 - x_3}{\phi_1 - \phi_3} , \tag{6.3}$$

$$C_v = \frac{\sigma_x}{\bar{X}} . \tag{6.4}$$

For the log-normal distribution, quantiles are again determined for the three probabilities selected, viz. $P_1 = 5\,\%$, $P_2 = 50\,\%$, and $P_3 = 95\,\%$.

$$\log\left(x_p - x_0\right) = \log\left(x_2 - x_0\right) + \sigma_u\,\phi_p(P, 0)\,, \tag{6.5}$$

Parameters x_0 and σ_u are computed from the following relationships:

$$x_0 = \frac{x_1 x_3 - x_2^2}{x_1 + x_3 - 2x_2}\,, \tag{6.6}$$

$$\sigma_u = 0.304 \log \frac{x_1 - x_0}{x_3 - x_0}\,, \tag{6.7}$$

σ_u being the standard deviation of substitute variable u, given by the following relationship,

$$u = \log\left(x - x_0\right).$$

The function $\phi_p(P, 0)$ expresses the theoretical (standardized) ordinates of the curve of transgression for $C_s = 0$. For selected ordinates the reader is referred to Table 15.

The parameters of the log-normal distribution can be estimated in a way analogous to that applied to the Pearson distribution. The index of skewness, S, is again computed first using equation (6.1) and then the coefficient of asymmetry, C_s, ascertained with its help. For that distribution relationship $S = f(C_s)$ can however be found listed in Table 14.

The other parameters, $\bar{X}$ and σ_x, or C_v, are again determined with the help of equations (6.2), (6.3), (6.4); the theoretical quantiles Φ_1, Φ_2, Φ_3 must however be read from Table 14.

The quantiles method can also be applied to the estimation of the parameters of other distributions. The literature, e. g. [30], presents the estimation of parameters of the double exponential Gumbel distribution, which is frequently applied to the analysis of the maximum terms of random samples. This distribution is characterized by the probability density (frequency curve)

$$\varphi(x) = \mathrm{e}^{-z}\,\mathrm{e}^{-\mathrm{e}^{-z}}\,, \tag{6.8}$$

where z denotes the standardized deviation from the mode; it is a function of random variable x according to the following relationship:

$$z = \frac{1}{0.779\,7\,\sigma}\left(x - \bar{x} + 0.450\,\sigma\right), \tag{6.9}$$

TABLE 14. Theoretical standardized quantiles of log-normal distribution

C_s	$\Phi(P, C_2)$		S	C_s	$\Phi(P, C_s)$		S
	Φ_5–Φ_{95}	Φ_{50}			Φ_5–Φ_{95}	Φ_{50}	
0.0	3.29	0.00	0.00	2.0	2.99	−0.24	0.42
0.1	3.29	−0.02	0.03	2.1	2.97	−0.24	0.44
0.2	3.29	−0.04	0.06	2.2	2.95	−0.25	0.45
0.3	3.28	−0.06	0.09	2.3	2.92	−0.25	0.46
0.4	3.28	−0.07	0.11	2.4	2.90	−0.26	0.48
0.5	3.26	−0.09	0.14	2.5	2.88	−0.26	0.49
0.6	3.25	−0.10	0.16	2.6	2.86	−0.26	0.50
0.7	3.24	−0.11	0.19	2.7	2.84	−0.27	0.51
0.8	3.22	−0.13	0.21	2.8	2.82	−0.27	0.51
0.9	3.21	−0.14	0.23	2.9	2.81	−0.27	0.52
1.0	3.19	−0.15	0.25	3.0	2.78	−0.28	0.53
1.1	3.17	−0.16	0.27	3.2	2.74	−0.28	0.55
1.2	3.16	−0.17	0.29	3.4	2.71	−0.29	0.56
1.3	3.14	−0.18	0.31	3.6	2.67	−0.29	0.57
1.4	3.11	−0.19	0.33	3.8	2.64	−0.29	0.58
1.5	3.10	−0.20	0.35	4.0	2.60	−0.29	0.59
1.6	3.07	−0.21	0.37	4.5	2.53	−0.30	0.62
1.7	3.05	−0.22	0.38	5.0	2.45	−0.30	0.64
1.8	3.03	−0.22	0.39				
1.9	3.01	−0.23	0.49				

TABLE 15. Quantiles of standardized normal distribution

P (%)	$\Phi_p(P, 0)$	P (%)	$\Phi_p(P, 0)$
0.1	3.09	80	−0.84
1	2.33	90	−1.28
5	1.64	95	−1.64
10	1.28	99	−2.33
20	0.84	99.9	−3.09
50	0.00		

or its general form:

$$z = a(x - q)\,. \tag{6.10}$$

Since the Gumbel distribution is asymptotic, equation (6.9) holds only for $n \to \infty$. With a more limited size of the set, the parameters of equation (6.10) must be corrected in accordance with the relations

$$a = \frac{s_{zn}}{s_x}\,, \tag{6.11}$$

$$q = \bar{x} + \bar{z}_n \frac{1}{a}\,. \tag{6.12}$$

The values of s_{zn} and $\bar{z}_n$ for various ranges of n are listed in Table 16.

TABLE 16. Parameters of the Gumbel distribution

n	$\bar{z}_n$	s_{zn}	n	$\bar{z}_n$	s_{zn}
20	0.524	1.063	54	0.550	1.167
22	0.527	1.076	56	0.551	1.170
24	0.530	1.086	58	0.552	1.172
26	0.532	1.096	60	0.552	1.175
28	0.534	1.105	65	0.554	1.180
30	0.536	1.112	70	0.555	1.185
32	0.538	1.119	75	0.556	1.190
34	0.540	1.126	80	0.557	1.194
36	0.541	1.131	85	0.558	1.197
38	0.542	1.136	90	0.559	1.201
40	0.544	1.141	95	0.559	1.204
42	0.545	1.146	100	0.560	1.206
44	0.546	1.150	500	0.572	1.259
46	0.547	1.154	1 000	0.574	1.269
48	0.548	1.157	∞	0.577	1.282
50	0.548	1.161			
52	0.549	1.164			

Since the equation of Gumbel's distribution (6.10) contains only two parameters, a and q, their estimation involves selecting two quantiles, x_1 (with $P_1 = 5\%$) and x_2 (with $P_2 = 95\%$).

It can be proved that

$$a = \frac{4.067}{x_1 - x_2}\,, \tag{6.13}$$

$$q = 0.27x_1 + 0.73x_2\,. \tag{6.14}$$

The ordinates x_p of the theoretical curve of transgression for various P's can be calculated from the equation

$$x_p = q + \frac{1}{a} z_p, \tag{6.15}$$

where z_p stands for the standardized deviation from the mode listed in Table 17.

TABLE 17. Normalized deviation from the mode of the Gumbel distribution

P	z_p	$\Phi(P)$	P	z_p	$\Phi(P)$
0.1	6.907	4.94	80	−0.476	−0.82
1	4.600	3.14	90	−0.834	−1.10
5	2.970	1.87	95	−1.097	−1.31
10	2.250	1.30	99	−1.527	−1.64
20	1.500	0.72	99.9	−1.933	−1.96
50	0.367	−0.16			

The ordinates x_p of the theoretical curve of transgression can also be computed from the following general equation:

$$x_p = \bar{x}[1 + C_v \, \Phi(P, C_s)] = \bar{x} + \sigma_x \, \Phi(P, C_s) \tag{6.16}$$

into which the values from Table 17 pertaining to Gumbel's distribution must in this case be substituted for $\Phi(P, C_s)$.

Parameters $\bar{x}$ and σ_x in equation (6.16) can be computed from the equations

$$\bar{x} = 0.412x_1 + 0.588x_2 , \tag{6.17}$$

$$\sigma_x = 0.315(x_1 - x_2) . \tag{6.18}$$

In order to justify the application of the quantiles method to water engineering, the drawbacks of the moments method, particularly the biassed estimators, are very often adduced in the water-engineering literature. The quantiles method is therefore sometimes preferred in spite of the fact that the properties of its estimators have so far not been studied in detail. In practice, this method is being widely used, owing particularly to its clarity and its computational simplicity.

From the point of view of the theory of estimation, it is the principle of the quantiles method itself that is the most important; from this principle it follows that the unknown characteristics[*)] of the sample ($\bar{X}$, σ, or C_v, C_s) are sought

[*)] We prefer using the term "characteristics" in this case, because no estimation of the parameters of the population is involved.

that will satisfy the condition that the theoretical line of transgression defined by these characteristics should cross the empirical quantiles selected. This theoretical procedure thus contributes to the theoretical line fitting closely the empirical line of transgression of the given sample, thus also to the detection of a fitting sample distribution. As already mentioned above, a most important question to be asked is whether the characteristics derived in this way have the weight of the unbiassed parameters of the population with its different types of distribution.

Our research was aimed at answering this question. Since analytic procedures had proved insufficient for the solution to be achieved, we again turned to modelling, an approach that we had already tested using the moments method and the maximum likelihood method. We generated 500 random samples of various length from 10 000-term random sequences. The quantiles method was used to estimate the characteristics of each sample, the whole 500-element sets of which were statistically further processed and the results were compared with the properties of the parameters of the given sequences. The detailed solution enabled us to satisfactorily assess the properties of the estimates arrived at using the quantiles method. (Comp. the next Section, 6.2).

This assessment however did not give any answer to another fundamental question, viz. whether the probability properties of parameter estimation can be upgraded by a more "reasonable" choice of the quantiles. Such research has already been started in Czechoslovakia, and it can be stated that, for instance, the introduction of a larger number of quantiles into the estimation of parameters of a distribution (as compared with the classical version of the quantiles method with only 2–3 quantiles considered) can substantially upgrade the properties of the estimators, particularly as far as their bias is concerned. Research has also shown that the coincidence of the theoretical and the empirical curves of transgression in the region of lower flows can be enhanced by the introduction of a new criterion, viz. the minimum sum of the squares of relative deviations applied to a larger number of quantiles. Research into these problems continues.

6.2 Properties of parameter estimates of populations with various probability distributions

6.2.1 Properties of parameter estimates of a population with Pearson's IIIrd type distribution

Research has arrived at new and useful results enriching the existing knowledge concerning the qualities of the hitherto held, mostly optimistic, views of the possibility of applying the method to the engineering practice.

The following are the five principal results:

1. Parameter estimation from a single random sample, making use of the quantiles method, can lead to considerable random errors, as in the case of the moments method. The values of parameters can only be approximated by the mean values of the set of sample characteristics. These can in some cases be less biassed than in the case of the moments method, and estimates can even be on the safe side, i. e. the expected values of the characteristics can exceed the parameters.

2. The advantage of the quantiles method is that with the Pearson distribution it substantially upgrades the estimation of skewness for all the lengths of the samples under examination. Its expected values may be biassed in either direction; the systematic errors are however markedly less grave (sometimes even negligible) than with the moments method. And for longer samples the estimates also converge towards parameters more rapidly than in the case of the moments method.

3. It is however a drawback of the quantiles method that with the Pearson distribution it biasses the estimates of both the coefficients of variation (as in the case of the moments method) and the means, the expected values of which are unbiassed with the moments method. With the models under examination the expected values of the means and the coefficients of variation were always higher than the parameters (the means by as much as 20 %, the coefficients of variation by as much as 13 %). As far as the application to culminating flows is concerned, such

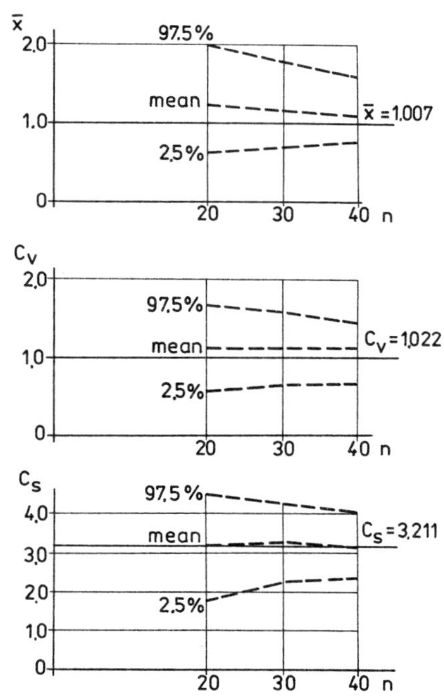

Fig. 33. Confidence intervals of sample means and coefficients of variation and asymmetry in a 10 000-element random sequence with Pearson's IIIrd type distribution obtained using the quantiles method. Inputs: $\bar{X} = 1.00$, $C_v = 1.00$, $C_s = 3.00$; Outputs: $\bar{X} = 1.007$, $C_v = 1.022$, $C_s = 3.211$.

estimates may be on the safe side (giving higher values), but the adequacy of that safety would have to be analyzed.

A typical example of these relationships is presented in Fig. 33, showing the curve of the expected values of the means and the coefficients of variation one-sidedly biassed above the parameters, and the curve of the insignificant systematic errors of the coefficient of asymmetry.

It is obvious that these qualities of the method can become a considerable drawback, the observation of the given profile available being often of a practically rather restricted character: the magnitude of the higher estimates and their adequate measure can neither be proved nor corrected without a simulation model being constructed. With the Pearson distribution the application of the method thus appears to be merely a suitable and supplementary methodological means of checking on the unbiassed skewness estimated with the help of other methods. For the estimation of the unbiassed mean and the coefficient of variation the moments method, with the bias corrected, seems to be the more suitable*).

4. Interesting probability properties are exhibited by the quantiles with the probability of transgression equal to 5 % and 95 %. The calculation of their statistical characteristics in the sets of 500 samples showed that the values of quantiles (particularly the left quantile) could vary within relatively wide limits and that they were therefore the cause of considerable random errors in parameter estimation. This fact also accounts for the quality of the quantiles method mentioned under item 1.

5. With the quantiles method, the length of the sample often exerts less effect upon the expected values of characteristics than with the moments method (particularly as far as the less biassed estimators are concerned). The length of the samples is more influential as far as the variance and the skewness of the parameters estimated are concerned. Variance (measured by the coefficient of variation of the means, the coefficients of variation and asymmetry) decreases with increasing length of the samples; with the skewness (measured by the coefficient of asymmetry) of these three characteristics this property manifests itself as a tendency allowing of some exceptions. The well-known fact that the safety of parameter estimation rises with lager samples is thus confirmed.

*) According to the final report of the Czechoslovak Hydrometeorological Institute [45] the Soviet standard [99] does not allow using the quantiles method for the triparametric gamma (Kritskiĭ–Menkeľ) distribution.

6.2.2 *Properties of parameter estimates of a population with log-normal distribution*

As in the preceding case, research has brought new and interesting results, which essentially change the hitherto held views concerning the application of the quantiles method to practice.

The main outcomes are as follows:

1. As with the Pearson distribution, estimation from a single sample can be burdened with considerable random errors. The mean values of the characteristics are more dependable, but in the given case these estimates are less adequate than with the Pearson distribution.

2. The estimation of the individual parameters is generally unsatisfactory. This estimation involves considerable systematic errors both with the coefficient of asymmetry and with the shorter samples. The methodological means of correcting this bias are however still lacking.

3. The quality of the estimates of the means and the coefficients of variation has markedly declined, all the expected values of the means and the coefficients of variation being lower as compared with the parameters (by up to 25 %) and totally inapplicable without being corrected. Similar risks are involved in the estimates of C_v, which are invariably biassed below the parameters.

This study has again led to the principal conclusion that for the estimation of parameters with triparametric log-normal distribution, the moments method with its bias compensated is the more suitable.

7
Analysis of time series, and their mathematical modelling

7.1 Fundamental problems of the analysis of time series

At present, the analysis of time series is one of the most significant dispciplines of mathematical statistics. It is the aim of this analysis to detect the probability properties of chronologically arranged data and to construct the corresponding mathematical model. The construction of the optimum fitting mathematical model depends to a great extent upon the estimation of the parameters. The analysis of time series is thus closely linked with the theory of estimation.

The data forming a time series are not only those supplied by various branches of technology, but also by physics, economics and other natural and social sciences. A time series can well be conceived of as a concrete materialization of a random sequence, the generation of which on a computer makes it possible to understand better the properties of the process under examination at other periods of time (viz. pseudo-chronological).

And moreover, the construction of the model facilitates forecasts of the development of the variables to be examined. The construction of the model makes it also possible to control and to optimize the operation of the given system by choosing the most suitable input parameters and initial conditions.

The analysis of time series and the construction of their models are of vital importance to hydrology and the other areas of water management. The mathematical models of various types of hydrological series enable generation of synthetic series, on the basis of which a number of important hydrological tasks can successfully be tackled. In the field of application there is increasing use, for instance, of synthetic flow series in the solution of the complex problems of reservoir design and systems of water management.

The analysis of time series was under the influence of deterministic approaches until about the first quarter of this century. The development of this discipline was greatly promoted by the work of Yule and Slutski as well as by the revealment of the possibility of using the models of moving totals of white noise for the generation of series with mutually dependent terms. At present, extraordinary attention is being paid to stochastic models of time series and various methods of smoothing time series. From the point of view of application the Box–Jenkins approach is of considerable importance.

Literature shows that modelling hydrological time series has already had a long tradition starting at the beginning of the present century. A survey of the historical development of that important discipline is given in the monograph by Salas, Delleur, Yevjevich and Lane [100].

The analysis of time series is linked to numerous problems, particularly in hydrology. The fundamental problems include above all the dependability of the observation of the hydrological variables and the length of that observation, which is of course related to the problem of estimating the distribution and its parameters. These problems are in turn related to the type and the derivation of the mathematical model as a basis for the generation of synthetic series.

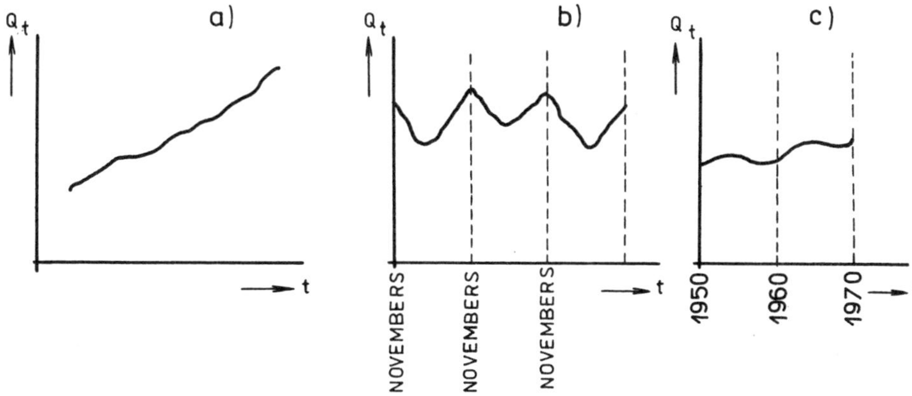

Fig. 34. Components of time series: a) trend component, b) seasonal component, c) cyclic component.

One of the fundamental approaches of the analysis of time series is the decomposition of these series into the following components: the trend, the seasonal component, the cyclical component and the residual (random) component (see the schematic diagram in Fig. 34). Decomposition of time series is invariably undertaken with the aim of finding a behaviour of the given series which is more regular (more deterministic) than the behaviour of the original series. Whereas the first three components – the trend, the seasonal and the cyclical components – can be viewed as kinds of function of time, the residual component represents a sequence of random variables.

The trend expresses long-term changes in the average behaviour of a time series (e. g. long-term rise or decline). The seasonal component reflects the periodically repeated annual changes in a time series, and the description of the seasonal component involves relatively frequent measurement. For some of the water-engineering tasks the expression of the seasonal component on the basis of the average monthly flows will prove insufficient and it then becomes necessary to use a shorter time interval. Admittedly, the construction of a better-fitting mathematical model will thus be more difficult.

The cyclical component, which has the character of a long-term non-stationary fluctuation about the trend (see Fig. 34), is invariably the most difficult to describe. Expressing it reliably is conditional upon long-term observation (measurement). This is also mostly linked with the difficult detection of the causes of the rise of the cyclical component. Its elimination is also hard to achieve if its character undergoes temporal changes.

The residual components remain in the time series after the trend and the seasonal and cyclical components have been separated. They reflect the random fluctuations in the behaviour of the time series which lack systematic character. It is usually assumed that the residual component represents white noise, mostly with normal distribution. This assumption need however not be always fulfilled. This is why more complex models of this component are sometimes constructed.

The decomposition of the time series can either be of an additive type, with the values of the terms of the series generated as a total of all the components, the components being considered in actual absolute values; or of a multiplicative type with the values of the terms of the series generated in the form of the product of all the components, only the trend components being considered in absolute values, the others in relative values, and dimensionless.

The decomposition of the time series usually assumes the individual observations to be uncorrelated variables. The Box–Jenkins approach, by contrast, makes it possible for the residual component to be taken as a correlated (dependent) random variable. This approach to the generation of dependence is quite frequent in Box–Jenkins' methodology and use is therefore made of correlation analysis. For the principles of this methodology the reader is referred to Box–Jenkins [14].

Among the basic models of that methodology belong the so-called models of moving averages of a certain q order, denoted as MA(q) models, the AR (p) autoregression models, and the ARMA models of p, q order. The Box–Jenkins models are more flexible than the decomposition methods*): they make it possible to model the stochastic seasonal component, and also the non-stationary tendencies in the time series.

The spectral analysis of these series is one of the important tasks of the analysis of time series. Spectral analysis views the time series as a sum of the sine and cosine components of various amplitudes and frequencies. Its aim is to give an insight into the inner structure of a given series, i. e. into the representation

*) By flexibility we understand the ability of the model to adapt itself to the changes in the time series.

of the individual frequencies in the series (into the so-called spectrum of the series), and to estimate for the coefficients of the periodic component. The problems of spectral analysis and the theoretical aspects of estimation of spectral densities from periodograms are receiving quite a lot of attention world wide.

Another important task of the analysis of time series, as well as the mathematical models linked with this analysis, is the construction of the forecasts of the future behaviour of the variable under examination. Dependable answers are of great importance to various branches of national economy, and the development of prognostic methods is thus a topical and an increasingly pressing task of contemporary research.

The forecasts can be based on both qualitative and quantitative methods. The qualitative methods are particularly employed in cases where the forecasts cannot be built on a past (historical) series of the values observed. Such qualitative methods rely upon the views and the claims of experts, they can therefore be burdened with subjective errors. This applies for instance to the Delphi method, according to which experts repeatedly voice their comments on a given prognostic problem (for instance, the development of science, technology, production) until a common view is arrived at as the final product of the method.

The quantitative methods start with a detailed statistical analysis of the given series of the values observed, on the basis of which a prognostic model is derived. This is practically tantamount to an extrapolation of the future from the contemporary values of the terms of the sequence. The reliability of the prognoses based upon these quantitative methods depends upon a number of factors: the probability properties of the given series of the values observed (particularly the autocorrelative and spectral properties) are of particular importance in this respect, as well as the length of the series, the length of the forecast (viz. the time remoteness of the value forecast from the present), the character of the given series (in the form of average annual, monthly or daily values), and the required precision of the forecast itself.

The quality of the forecasts is assessed by various indicators. Use is often made of the sum of squared errors (SSE), the mean squared error (MSE), or the mean absolute deviation (MAD).

The derivation of the mathematical models from a time-limited observation of variables is one of the most important, but also the most difficult, undertakings involved in the solution of hydrological water management problems. We must realize that the creation of a model representation of the reality with the help of a relatively exact methodological apparatus in no way eliminates the inexactitude and vagueness of the original assumptions concerning the properties of the original real (historical) series. And the dependability of the observation itself is a no less serious problem.

These problems are linked with a certain dilemma: in order to gain knowledge of the properties of hydrological reality we construct models of that reality, which are expected to help us discern these properties. However, a model describing adequately the reality can only be built upon good knowledge of the properties of that reality.

The way out of this dilemma lies in a suitable iterative method of modelling, in the transition from simpler to more complex models.

7.2 Basic models of time series

As already mentioned in the preceding paragraph, the basic models of Box–Jenkin's methodology are: the model of moving sums, the autoregression model and mixed models. The present section will deal with the fundamental properties of these models.

The process of moving averages of q order, MA(q), is invariably given the following form:

$$y_t = \varepsilon_t + \Psi_1 \varepsilon_{t-1} + \ldots + \Psi_q \varepsilon_{t-q} , \tag{7.1}$$

where ε_t denotes white noise and Ψ_j the parameters.

Using the symbols of the operator of backward displacement B, $By_t = y_{t-1}$, process (7.1) can be rewritten in the following form:

$$y_t = \Psi(B)\, \varepsilon_t , \tag{7.2}$$

where

$$\Psi(B) = 1 + \sum_{j=1}^{q} \Psi_j B^j \tag{7.3}$$

is the operator of moving averages.

The mean value of this process is zero, and the variance σ_y^2 equals

$$\sigma_y^2 = (1 + \Psi_1^2 + \ldots + \Psi_q^2)\, \sigma_\varepsilon^2 . \tag{7.4}$$

The autocorrelation function ϱ_k has the following form*):

$$\varrho_k = \frac{\Psi_k + \Psi_1 \Psi_{k+1} + \ldots + \Psi_{q-k} \Psi_q}{1 + \Psi_1^2 + \ldots + \Psi_q^2} \quad \text{for } k = 1, \ldots, q ,$$

$$\varrho_k = 0 \text{ for } k > q . \tag{7.5}$$

*) In order to simplify the notation of correlation functions we denote time remoteness as an index, not as a bracketed argument.

Apart from the autocorrelation function, the Box–Jenkins method defines the so-called partial autocorrelation functions ϱ_{kk} expressing partial correlation coefficients y_t and y_{t+k} with fixed values of $y_{t+1}, \ldots, y_{t+k-1}$. The partial autocorrelation function can be computed using the following expression [14, 25]

$$\varrho_{kk} = \frac{|\mathbf{P}_k^*|}{|\mathbf{P}_k|}, \tag{7.6}$$

where | | stands for the determinant of the matrix and $\mathbf{P}_k$ for the matrix of autocorrelations in the following form,

$$\mathbf{P}_k = \begin{bmatrix} 1, \varrho_1, & \ldots, \varrho_{k-1} \\ \varrho_1, 1, & \ldots, \varrho_{k-2} \\ \ldots & \\ \varrho_{k-1}, \varrho_{k-2}, \ldots, & 1 \end{bmatrix} \tag{7.7}$$

and matrix $\mathbf{P}_k^*$ is generated from matrix $\mathbf{P}_k$ by modification of the last column:

$$\mathbf{P}_k^* = \begin{bmatrix} 1, \varrho_1, & \ldots, \varrho_1 \\ \varrho_1, 1, & \ldots, \varrho_2 \\ \ldots & \\ \varrho_{k-1}, \varrho_{k-2}, \ldots, & \varrho_k \end{bmatrix} \tag{7.8}$$

It thus holds that, for instance,

$$\varrho_{11} = \varrho_1 , \tag{7.9}$$

$$\varrho_{22} = \frac{\begin{vmatrix} 1, & \varrho_1 \\ \varrho_1, & \varrho_2 \end{vmatrix}}{\begin{vmatrix} 1, & \varrho_1 \\ \varrho_1, & 1 \end{vmatrix}} = \frac{\varrho_2 - \varrho_1^2}{1 - \varrho_1^2} . \tag{7.10}$$

The autocorrelation function and the partial autocorrelation function coincide at the point $k = 1$.

The estimates r_{kk} of the partial autocorrelation function ϱ_{kk} can be computed using recurrent expression $r_{11} = r_1$,

$$r_{kk} = \frac{r_k - \sum_{j=1}^{k-1} r_{k-1,j}\, r_{k-1}}{1 - \sum_{j=1}^{k-1} r_{k-1,j}\, r_j} \quad \text{for } k > 1. \tag{7.11}$$

where

$$r_{k,j} = r_{k-1,j} - r_{kk}\, r_{k-1,k-j} \quad \text{for} \quad j = 1, 2, \dots, k-1\,. \tag{7.12}$$

Similarly as with the autocorrelation function ϱ_k (comp. Section 3.3), testing can be applied to the zero values of partial autocorrelation function ϱ_{kk}, using the Quenouille approximation [93] for the standard deviation of estimate r_{kk}: if $\varrho_{kk} = 0$ for $k > k_0$, then

$$\sigma(r_{kk}) \cong \sqrt{\frac{1}{n}}\,, \qquad k > k_0\,. \tag{7.13}$$

In the test itself use is made of a double of standard deviation $\sigma(r_{kk})$, as in the case of the autocorrelation function (comp. Section 3.3).

The autocorrelation functions and partial autocorrelation functions of processes MA(1) and MA(2) can be cited as examples in this respect.

The autocorrelation function ϱ_k of process MA(1), given in the following form,

$$y_t = \varepsilon_t + \Psi_1\, \varepsilon_{t-1}\,, \tag{7.14}$$

equals

$$\varrho_1 = \frac{\Psi_1}{1 + \Psi_1^2}\,, \tag{7.15}$$

$\varrho_k = 0$ for $k > 1$.

For the partial autocorrelation function the following expression can be derived [6, 25],

$$\varrho_{kk} = \frac{(-1)^{k-1}\, \Psi_1^k (1 - \Psi_1^2)}{1 - \Psi_1^{2(k+1)}}\,, \tag{7.16}$$

this function being limited by the geometrically declining sequence

$$|\varrho_{kk}| < |\Psi_1|^k\,. \tag{7.17}$$

The MA(2) process can be expressed in the following form:

$$y_t = \varepsilon_t + \Psi_1 \varepsilon_{t-1} + \Psi_2\, \varepsilon_{t-2}\,. \tag{7.18}$$

The ordinates of the autocorrelation function of this process acquire the following values:

$$\varrho_1 = \frac{\Psi_1(1 + \Psi_2)}{1 + \Psi_1^2 + \Psi_2^2},$$
$$\varrho_2 = \frac{\Psi_2}{1 + \Psi_1^2 + \Psi_2^2},$$
$$\varrho_k = 0 \quad \text{for} \quad k > 2\,. \tag{7.19}$$

The computation of the partial autocorrelation function for process MA(2) is much more complex (this function either being limited by the geometrically declining sequence, or having the form of a sinusoid with geometrically declining amplitude).

The process of autoregression of p order is denoted as AR(p) and rewritten as

$$y_t = \varphi_1 y_{t-1} + \ldots + \varphi_p\, y_{t-p} + \varepsilon_t\,, \tag{7.20}$$

or, using the symbolism of the operator of backward displacement B, as

$$\varphi(B)\, y_t = \varepsilon_t\,, \tag{7.21}$$

where

$$\varphi(B) = 1 - \sum_{j=1}^{p} \varphi_j\, B^j$$

is the autoregression operator.

Process AR(p) is stationary if all the roots of polynomial $\varphi(B)$ lie outside the unity-radius circle.

The mean value of stationary process AR(p) is zero and its autocorrelation function ϱ_k satisfies the system of finite difference equations

$$\varrho_k = \varphi_1\varrho_{k-1} + \varphi_2\varrho_{k-2} + \ldots + \varphi_p\varrho_{k-p}, \qquad k > 0\,. \tag{7.22}$$

These equations are derived so that relationship (7.20) is gradually multiplied by variables y_{t-k} for $k > 0$ and the expected values are calculated. Use is made of the relationship $E(y_{t-k}\,\varepsilon_t) = 0$ for $k > 0$. For the autocorrelation function of process AR(p) the following relationship can be arrived at:

$$\varrho_k = \alpha_1 G_1^{-k} + \ldots + \alpha_p G_p\,, \qquad \mathrm{k} \geqq 0\,, \tag{7.23}$$

where $\alpha_1, \ldots, \alpha_p$ are constants and $G_1, \ldots, G_p$ are roots of the polynomial $\varphi(B)$.

The parameters of process AR(p) can be computed from the values of its autocorrelation function with the help of the Yule–Walker system of equations consisting of relationships (7.22) for $k = 1, \dots, p$:

$$\left.\begin{aligned} \varrho_1 &= \varphi_1 + \varphi_2\varrho_1 + \dots + \varphi_p\varrho_{p-1}, \\ \varrho_2 &= \varphi_1\varrho_1 + \varphi_2 + \dots + \varphi_p\varrho_{p-2}, \\ &\vdots \\ \varrho_p &= \varphi_1\varrho_{p-1} + \varphi_2\varrho_{p-2} + \dots + \varphi_p. \end{aligned}\right\} \tag{7.24}$$

The solution of this system of linear equations will give us parameters $\varphi_1, \dots, \varphi_p$ expressed by values $\varrho_1, \dots, \varrho_p$ of the autocorrelation function.

Variance σ_y^2 of the process AR(p) is equal to

$$\sigma_y^2 = \frac{\sigma_\varepsilon^2}{1 - \varphi_1\varrho_1 - \dots - \varphi_p\varrho_p}. \tag{7.25}$$

Process AR(1) is the simplest case of process AR(p). In AR(1) the value of process y_t depends upon only a single preceding value y_{t-1} and the simultaneous value of white noise in accordance with the following relationship:

$$y_t = \varphi_1 y_{t-1} + \varepsilon_t \tag{7.26}$$

with the condition of stationarity $|\varphi_1| < 1$; with this condition holding, the process can be rewritten in the following form [25]:

$$y_t = \varepsilon_t + \varphi_1\varepsilon_{t-1} + \varphi_1^2\varepsilon_{t-2} + \dots. \tag{7.27}$$

The autocorrelation function of stationary process AR(1) has the following form:

$$\varrho_k = \varphi_1^k, \qquad k \geqq 0, \tag{7.28}$$

which is a geometrical sequence declining to zero in the absolute values. Since it holds that $\varphi_1 = \varrho_1$, relationship (7.28) can be given in the form

$$\varrho_k = \varrho_1^k. \tag{7.29}$$

Process AR(2) has the following form:

$$y_t = \varphi_1 y_{t-1} + \varphi_2 y_{t-2} + \varepsilon_t \tag{7.30}$$

with the conditions of stationarity [25]

$$\varphi_2 + \varphi_1 < 1, \qquad \varphi_2 - \varphi_1 < 1, \qquad -1 < \varphi_2 < 1, \tag{7.31}$$

which are diagrammed in Fig. 35.

If G_1 and G_2 are two different roots of the equation

$$1 - \varphi_1 B - \varphi_2 B^2 = 0\,, \tag{7.32}$$

then the autocorrelation function of process AR(2) has the following form:

$$\varrho_k = \frac{G_1^{-1}(1 - G_2^{-2})G_1^{-k} - G_2^{-1}(1 - G_1^{-2})G_2^{-k}}{(G_1^{-1} - G_2^{-1})(1 + G_1^{-1}G_2^{-1})}\,. \tag{7.33}$$

In the case of real roots G_1 and G_2 (with $\varphi_1^2 + 4\varphi_2 \geqq 0$) the function is a linear combination of two geometrically declining sequences, whereas in the case of complex conjugate roots (with $\varphi_1^2 + 4\varphi_2 < 0$) its curve is sinusoidal with the amplitude declining geometrically.

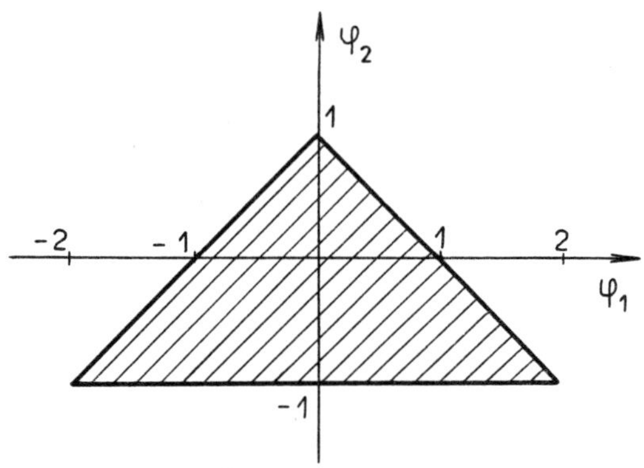

Fig. 35. Stationarity region of the AR(2) process.

The mixed process of orders p and q denoted as ARMA (p, q) is defined as

$$y_t = \varphi_1 y_{t-1} + \ldots + \varphi_p y_{t-p} + \varepsilon_t + \Psi_1 \varepsilon_{t-1} + \ldots + \Psi_q \varepsilon_{t-q}\,, \tag{7.34}$$

and using the symbolism of the operator of backward displacement B, in the following form:

$$\varphi(B) y_t = \Psi(B)\, \varepsilon_t\,. \tag{7.35}$$

It can be shown that the autocorrelation function of process ARMA(p, q) satisfies a similar system of finite difference equations as in the case of the autoregressive process, viz.

$$\varrho_k = \varphi_1 \varrho_{k-1} + \varphi_2 \varrho_{k-1} + \ldots + \varphi_p \varrho_{k-p}\,, \qquad k > q\,. \tag{7.36}$$

This system is derived analogously as in the case of process AR(p) and its solution has a form analogous to (7.23).

The classical Box–Jenkins methodology offers, apart from the basic MA(q), AR(p), ARMA(p, q) models, further possibilities of generating the more complex types of time series. In this context, mention can be made of the integrated mixed process ARIMA(p, d, q) making it possible to model stochastically the trend component, apart from the random fluctuations, of course. Besides the trends requiring stochastic modelling, the ARIMA models can also cope with deterministic trends.

The integrated mixed ARIMA(p, d, q) model is often given the following written form,

$$\varphi(G)\, w_t = \Psi(B)\, \varepsilon_t\,, \tag{7.37}$$

where

$$w_t = \Delta^d y_t \tag{7.38}$$

is the d-th difference of the process y_t modelled, and (7.37) is virtually the stationary ARMA(p, q) model of process w_t.

The first differences of series y_t are defined as

$$\Delta y_t = y_t - y_{t-1}\,, \tag{7.39}$$

the second differences as

$$\Delta^2 y_t = \Delta y_t - \Delta y_{t-1} = (y_t - y_{t-1}) - (y_{t-1} - y_{t-2}) = y_t - 2y_{t-1} + y_{t-2}\,, \tag{7.40}$$

and generally the difference of order k as

$$\Delta^k y_t = y_t - \binom{k}{1} y_{t-1} + \binom{k}{2} y_{t-2} - \dots + (-1)^k y_{t-k}\,. \tag{7.41}$$

The differential operator Δ can be expressed with the help of the operator of backward displacement B as

$$\Delta = 1 - B\,, \tag{7.42}$$

for

$$\Delta y_t = y_t - y_{t-1} = \left(1 - \frac{y_{t-1}}{y_t}\right) y_t = (1 - B) y_t\,. \tag{7.43}$$

We can thus formally write

$$\Delta^2 y_t = (1 - B)^2 \, y_t = \\ = (1 - 2B + B^2) y_t = y_t - 2y_{t-1} + y_{t-2} \,. \tag{7.44}$$

The ARIMA(p, d, q) model can thus be rewritten in the following form:

$$\varphi(B)(1 - B)^d y_t = \Psi(B) \varepsilon_t \,. \tag{7.45}$$

Building the ARIMA model invariably involves transforming suitably the original y_t series and differentiating it to series w_t, for which the ARMA(p, q) model is then constructed.

The classical models explain the behaviour of the time series solely on the basis of the given elements of the series itself (thus e. g. using the historical record of the series). The same also applies to the construction of the prediction of the future terms of the series. Another approach to the construction of the models is based upon the utilization of other time series, the behaviour of which is employed practically to explain the properties of the behaviour of the given series. The properties of variable y_t to be explained are thus given by the properties of explanatory variables x_t. This approach to the construction of the models can lead to quite a wide spectrum of regression models, since the variable y_t to be explained can be linked with a larger number of explanatory variables in time series x_t, u_t, v_t, ... Moreover, the dynamic of these models can also be based upon the utilization of the bonds between the variables variously displaced in time, which produces models with lagged explanatory variables. In the case of a single explanatory variable x_t, these models have the following general form:

$$y_t = \delta_0 x_t + \delta_1 x_{t-1} + \delta_2 x_{t-2} + \dots + u_t \,, \tag{7.46}$$

where coefficients δ_i, $i = 0, 1, \dots$ are the coefficients of the i-th time lag and u_t denotes the residual component.

The whole process of constructing the models of time series can invariably be divided into three principal phases. The first is the identification phase, the aim of which is to decide on a fitting type of model and to determine the order. The second is the phase in which the parameters of the model are estimated, and in the third phase the properties of the model constructed are finally assessed.

For the identification of the model a detailed analysis of the given time series is indispensable, particularly an analysis of its probability properties. An analysis of the stationarity of the series, its autocorrelation properties and the assessment of the type of seasonality are essential. These analyses are then made the basis for the selection of the concrete type of the mathematical model, and for the preliminary estimation of the parameters of that model.

The identification of the model can in some cases be a most difficult task, in which a suitable model is to be selected from several alternatives for the given time series. That is why objective and computer-friendly identification procedures are sought in order to substitute for the decision-making statistician, thus eliminating any subjective bias concerning the choice of the model.

The estimation of the order of the model can also be quite a numerically exacting task due to the fact that several points of view must be taken into account (viz. both the goodness of fit of the model and its size, the acquisition of the point estimate of the order of the model and its easy incorporation into a computer programme). At present, use is made for estimating the order of the model of the so-called penalizing functions, which penalize the choice of an excessive order of the model. We will deal with this problem in more detail in Section 11.3.

In the second phase, parameters of the models are sought, invariably with the help of the optimization methods, with a certain point of view selected as the criterion of optimality (e. g. the minimum sum of squared residual deviations). For optimum estimation of parameters use is often made of iterative procedures performed on large computers. Special literature draws attention to the fact that development in this field manifests a tendency towards full automation of the process of analyzing the given data and deriving the most fitting model.

The third phase of the construction of the model is a check upon its properties, and confirmation, or rejection, of the models's adequacy. If the model does not prove satisfactory, the whole procedure of its derivation must be repeated. In this phase, too, various statistical tests are used to ascertain the agreement between the properties of the given series and the properties of the model. When synthetic hydrological series are modelled, the usual requirement is for the statistical characteristics of the given (historical) series to agree with the parameters of the synthetic series, except for the random deviations at the 5 % level of significance.

Part II
Application of estimation theory to hydrology and water engineering

8
Parameter estimation of series of maximum flood flows

8.1 Fundamental problems of processing N-year flows

The problems of processing culminating flood flows were given considerable importance as early as at the time when probability theory had just begun to develop and when the methods of the theory of estimation had not yet been satisfactorily elaborated. The variable properties of hydrological regimes, a large number of factors affecting flood runoffs, the limited length of observation, and the related problems of estimating the law of probability distribution, as well as the extrapolation in the region of low probabilities of transgression – all these circumstances are the reason why finding the desired hydrological design quantities has always been one of the most difficult tasks in the processing of hydrological information.

With the development of knowledge in this field further complex problems gradually started to be investigated, such as for instance the probability properties of the flood flows and their genesis in the different seasons of the year (rain-induced and snow-induced floods), extreme runoffs from smaller river basins not subjected to measurement, the effect of historical floods on the parameters of a series of culminating flows, regional relationships of extreme runoffs.

At present, use is currently made of statistical and genetic methods for determining the design parameters of the flood waves, and empirical formulae have also been derived. These methods, as well as the conditions of their application, have been described in the already quite voluminous hydrological literature.

One of the most comprehensive and important works published in Czechoslovakia concerning the fundamental characteristics of hydrological phenomena

TABLE 18. Characteristics of maximum annual flows in a set of 250 stations in Czechoslovakia (according to [38])

Length of the periods examined	C_v	σ_{C_v} (%)
the shortest period of 25 years considered	0.4 to 1.0	18 to 29
the most frequent period of 30 years	0.3 to 1.2	15 to 30
the second most frequent period of 55 years	0.4 to 0.9	12 to 18
the period of 80 to 85 years	0.5 to 0.8	10 to 14

TABLE 19. Characteristics of maximum annual flows in a set of 250 stations in Czechoslovakia (according to [38])

Type of observation	C_s	C_v	σ_{C_s} (%)
minimum observation of 25 years	0.9 to 2.6	0.4 1.0	27 to 79 65 to 189
the most frequent 55-year observation	0.6 to 2.8	0.4 0.9	17 to 80 36 to 166
the longest observation series of 85 years	1.5 to 2.5	0.5 0.8	18 to 30 15 to 25

are the Hydrologické poměry ČSSR (Hydrological Regimes of the Czechoslovak Socialist Republic*) [51]). In this work, properties of the flood flows of Czechoslovak streams are dealt with in Part III Chapter 7, which presents the statistical characteristics of culminating flows in a set of 250 stations, their hydrological and geographic characteristics, as well as an analysis of the basic factors affecting runoffs.

---

*) In the year of publication the official name of the state was the Czechoslovak Socialist Republic (ČSSR).

From the point of view of the theory of estimation, and the justification of the application of that theory, the statistical characteristics of the culminating flows must be regarded as the most valuable. They were computed with the help of the moments method, or the quantiles method,*) using the shortest, 25-year, series and the longest, 115-term, series (at Děčín on the Elbe) of the set. The most significant results are given in Tables 18 and 19. The lowest C_v value, 0.23, was recorded at Komárno, and the highest value, 1.30, at Husinec. The most frequent C_v value ranged between 0.50 and 0.69 (with the median equal to 0.6). For the coefficient of asymmetry the lowest value, 0.03, was recorded at Michalovce, and the highest, 3.2, at Spálov.

From the results quoted it is evident that the real series of maximum annual flows exhibit considerable fluctuation and skewness, and that the length of observation was in several cases rather limited. These properties of the culminating flow series substantiate the necessity to estimate for their unbiassed parameters. The greatest attention should be given to the distribution of probability and to the ascertainment of the systematic errors with its asymmetrical types. Only in this way can dependable values of the *N*-year flows be approximated.

The application of the theory of estimation can be traced back to 1977, when basic material was being processed for the research project entitled The Complex Solution of the Water Engineering Problems of the North-Bohemian Lignite Basin and the Related Problems of the Protection of the Environment. It was then that studies of the laws of the flood regimes of the smaller streams started appearing, based upon the theory of estimation and the application of simulation models of hydrological processes [19]. The research led to new knowledge concerning the behaviour of the culminating flows and their sample characteristics.

It particularly turned out that mechanical application of the hitherto current methodological procedures based upon the assumption of the representativeness of a single short sample could on the average lead to systematic underestimation of *N*-year flows. The research also confirmed the fact that for smaller streams the estimation of the parameters of their culminating flows is of particular importance, because their regimes are characterized by high variability and skewness of distribution (the differences in the *N*-year maximum flows amounting to as much as several hundred percent).

Figure 36 presents a characteristic example of a period of chronologically ordered culminating flows in a modelled 10 000-years series with the parameters estimated. The example shows that under the extreme conditions of the smaller streams, catastrophic flows can occur by sheer chance after a calm period of

*) As in the preceding works, no corrections were considered of the characteristics calculated as far as the systematic errors are concerned; at that time the required relationships between characteristics and parameters had not yet been formulated.

several decades. A proposal for concrete antiflood measures to be adopted based upon short-term observation and underestimating outcomes of the laws of statistics can be extraordinarily risky, and could lead to serious economic consequences.

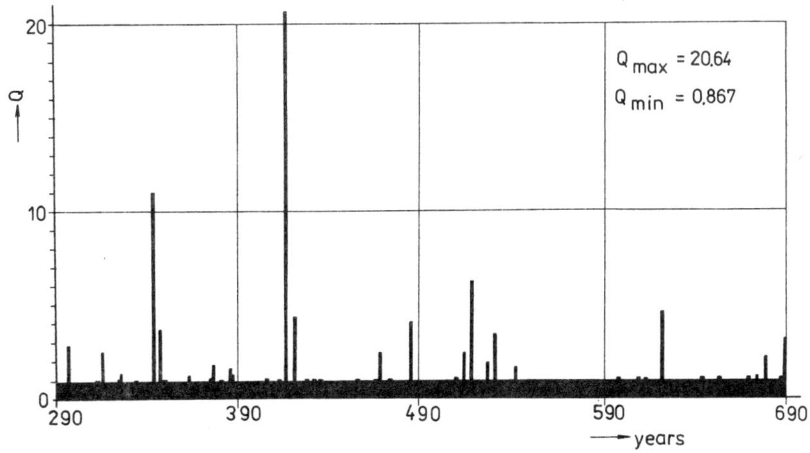

Fig. 36. Characteristic section in a 10 000-year random series of maximum annual flows with Pearson's IIIrd type distribution
Inputs: $\bar{Q} = 1$, $C_v = 0.8$, $C_s = 12$.

The processing of the N-year maximum flows has received much attention at the Czech Hydrometeorological Institute in Prague. Its report [45] contains a summary assessment of the latest achievements in the application of the estimation theory to bulk processing of the culminating flows. The recommendation of the most suitable types of theoretical distributions as well as the methods of estimating their parameters is most important. As far as practical application is concerned, it is suggested that use should primarily be made of Pearson's IIIrd type, triparametric log-normal, and logarithmic Pearson's IIIrd type distributions. Gamma distribution is not recommended in view of the fact that in the region of the lower probabilities of transgression it will lead to results analogous to those of the computationally simpler triparametric log-normal distribution.

In agreement with the results of the research conducted by the Department of Hydrotechnology of the Czech Technical University in Prague, the moments method, with the systematic bias involved in the estimation of the coefficients of variation and asymmetry corrected, is recommended for bulk processing of the N-year flows for all the types of theoretical distribution quoted above. For reasons mentioned in Part I of this book, neither the maximum likelihood method nor the quantiles method is reccommended for bulk data processing.

The automatic bulk processing of the series of culminating flows and the determination of the N-year maximum flows revealed the necessity to upgrade the efficiency of the moments method and to convert the graphical relationships concerning the estimation of the unbiassed parameters to analytic form, which is of course more computer-friendly. The conclusions formulated in the foreign literature available to us [12, 98] were of course taken into account, as well as the results of our own comprehensive research [81, 82]. In view of the general importance of the solution, i. e. also for parameter estimation of other types of series and their mathematical modelling, this subject is treated separately in Section 11.1.

The Czech Hydrometeorological Institute has completed a draft for complex automatic processing of the N-year maximum flows [45, 46] corresponding to the contemporary level of knowledge supplied by the theory of estimation achieved both in Czechoslovakia and abroad. The programme is a valuable outcome of the long research conducted by the Department of Hydrotechnology of the Czech Technical University in Prague in close cooperation with the Czech Hydrometeorological Institute in Prague.

Apart from the programme itself, aids have been prepared to facilitate the computation of the design variables in cases where a computer is not available.

As far as the probability properties of the culminating flow series are concerned, a most important problem is the determination of the design N-year flows with due account taken of historical floods.

In the literature on this subject [30, 45] we find expressions for the estimation of the mean values of the coefficients of variation and asymmetry of the culmination flow series, with the occurrence of historical floods duly considered. These problems have recently been dealt with by Kašpárek, who in his study [42] gave an evaluation of the significance of the floods on the Litavka in the years 1872 to 1981 for the estimation of the N-year flows. In a number of variants their computations revealed that the effect of the historical floods on the determination of the magnitudes of the N-year flows could be most significant. A full-scale investigation and adequate processing of the data on extraordinary floods, whether they have occured only recently or in the past, could, in a number of cases, reduce the risk of estimating wrongly the design flows. Whenever anti-flood precautions are to be adopted, it is thus essential that these circumstances should be taken fully into account.

Despite the results achieved so far in the field of application of the theory of estimation, research must be continued and attention should be given to the problems that have so far remained unsolved, such as the problem of theoretical distributions of the flow series with historical floods; smoothing of the results achieved with the help of numerical procedures so as to make them applicable to the whole river-basin, with due account taken of its hydrological regularities; and determination of the N-year flows in the smaller river-basins where the required observations may be lacking. Apart from the culminating flows, more attention will have to be given to the shape and the volume of the flood waves,which should be regarded as basic information, besides the culminating flows, upon which the design the protective effect of storage reservoirs is based.

8.2 Probability properties of intervals between culminating flows

By "interval between culminating flows" we mean the interval at which the culminating flow selected repeats itself. In practice use is invariably made of its mean value. In our research we conceived of it as a random variable describable with the help of the respective statistical characteristics.

It was the aim of our research to investigate the fundamental probability

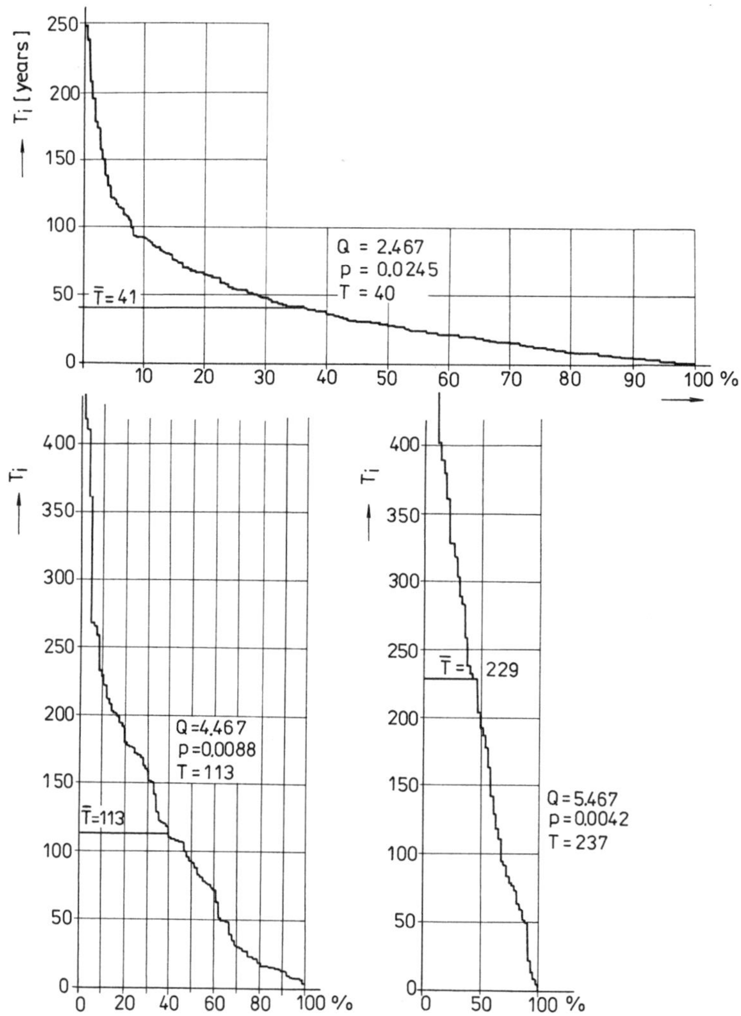

Fig. 37. Lines of transgression of all T_i times between selected maximum annual flows of a 10 000-year random series with Pearson's IIIrd type probability distribution
Inputs: $\bar{Q} = 1$, $C_v = 0.8$, $C_s = 12$;
Outputs: $\bar{Q} = 0.990$, $C_v = 0.682$, $C_s = 10.800$.

properties of these intervals and to explain more profoundly the relationships between the sample observation and the population.

Use was made of the modelled 10 000-year random series of culminating flows generated as absolutely random sequences with specified parameters. The printed output of the model were the lines of transgression of all the intervals of repetition covering the whole scope of the culminating flows.

Figure 37 presents examples of the lines of transgression of the intervals of repetition obtained from a 10 000-year random series with Pearson's type III distribution under the extreme conditions of a small (unwooded) river-basin, where the culminating flows could exhibit a high degree of fluctuation and skewness. The theoretical values of the average intervals of repetition, T, were calculated by using the following formula:

$$T = \frac{0.434\,29}{\log\,(1 - p)}, \tag{8.1}$$

where p is the probability of transgression.

Formula (8.1) can be derived from Poisson's law of distribution. The values of T were in all the cases compared with the expected values of the empirical lines of transgression.

From the examples presented in Fig. 37 it can be seen that it is fully justified to consider the intervals of repetition as random variables exhibiting relatively high variance. Thus, for instance, 113-year maximum flows ($Q = 4.467$) occured (i. e. were reached or exceeded) in one case within a 4-year period, another extreme being the period of 637 years for which that flow did not reappear.

Analogous properties are manifested by the curves of transgression of the intervals of repetition (Fig. 38), which correspond approximately to the regimes of small, partly wooded and sloping river-basins. For instance, in one case a 101-year flow was repeated in the next year but one, the contrary extreme being the period of 385 years, during which that climax did not reappear. It is also characteristic of the lines of transgression that with the N-year flow rising, the potential variance of the intervals of repetition grows quite rapidly. A 5.826 climax which repeats (i. e. it is reached or exceeded) in 196 years on average, may not occur for as long as 892 years; and a 9.326 climax repeated in 1000 years on average, may reappear in 115 to 3 460 years (in Fig. 38 these extremes have of course not been plotted).

The results achieved fully support the previous results of the research into the behaviour of the sample characteristics and their relationship to parameters. From the lines of transgression of the intervals between the climaxes it can be seen that in the shorter periods of observation, such as periods of several tens of years, no significant extreme flow may occur at all, leading to more dependable extrapolation of the lines of transgression into the region of the lower probabilities. The contrary case can however not be excluded either, viz. the occurrence of several extremes in a shorter period, which may lead to overestimation of the probability properties of the given phenomenon.

Whereas the systematic bias can relatively easily be eliminated using contemporary methods, the estimation of random errors is more difficult, because in view of the nature of the given hydrological phenomenon it may not be so easy to ascertain whether the series observed is representative or not. The checks on the representativeness of the culminating flow series will henceforward have to be based primarily upon the genetic and comparative methods.

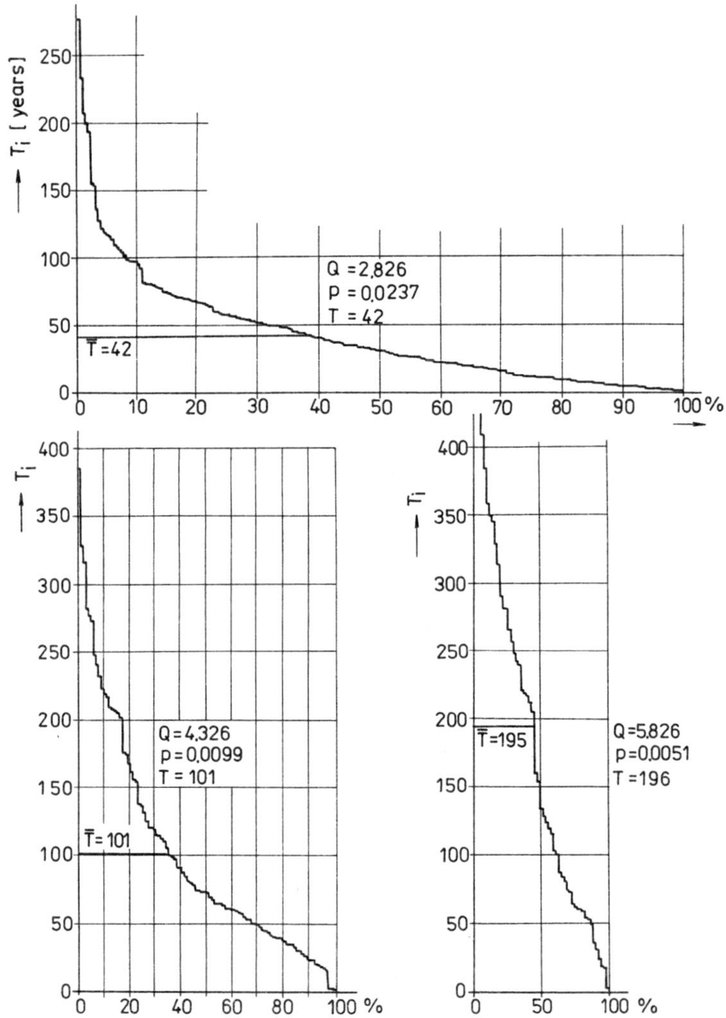

Fig. 38. Lines of transgression of all T_i times between selected maximum annual flows of a 10 000-year random series with Pearson's IIIrd type probability distribution
Inputs: $\bar{Q} = 1$, $C_v = 0.7$, $C_s = 8$;
Outputs: $\bar{Q} = 1.002$, $C_v = 0.717$, $C_s = 7.953$.

9
Estimation of parameters of average annual flow series

9.1 Estimation of parameters of probability distribution

In spite of the fact that the probability properties of the annual flow series may not be so extreme as the probability properties of the culminating flood flows, these properties have received attention ever since probability methods started being applied in water engineering. Research has concentrated on both the investigation of the possibilites of utilizing the periodic tendencies manifested by these series for long-range statistical forecasts of river runoffs and the design of storage reservoirs, the parameters of which are markedly dependent upon the properties of these series.

Contemporary water engineering literature has a large number of works dealing with the probabilistic and genetic properties of the average annual flow series. Mention should particularly be made of the works documenting the development of the probability methods of designing storage reservoirs starting in the early decades of the present century. It can well be claimed that the development of these methods has been dependent upon the level of knowledge of the properties of the flow series, upon the adequacy of the explanation of their causation, and upon the quality of their mathematical expression.

The early methods of designing water storage reservoirs were founded upon the simplest assumption of the absolute independence of the average annual flows (with zero autocorrelation function). The development then continued via the simple Markov chain up to the complex Markov chain, which is already quite capable of giving an adequate description of the internal structure of a given series.

The development of the design of water storage reservoirs has had a considerable effect upon mathematical modelling of the flow series, the development of which started in the mid-sixties. It is a substantial advantage of mathematical modelling that it enables simulation of relatively complex probability properties of the flow series, the introduction of which into analytical methods invariably proves to be fairly difficult.

Besides the development of the knowledge of the fundamental probability properties of the real flow series, the investigation of the historical development of the representativeness of these series has also proved to be of considerable interest. For more information on the principal stages of this development the reader is referred to Chapter 2 of the present monograph.

In the investigation of the representativeness of the average annual flow series and the estimation of long-term parameters, use has so far been made of various methods of comparative analysis – of both the properties of various samples of the same series and the properties of a suitable analogue.

The variability of the runoff of some of the Czechoslovak rivers has been examined in detail by Votruba and Brоža [116]. In their analyses these authors, too, made use of the sample statistics of various periods. The results of these analyses have led to valuable conclusions concerning the variability of the properties of the flow series in time and space, and they have lent considerable concreteness to the idea of the reliability of the individual values used in designing water storage reservoirs.

Several monographs by Bratránek can also be viewed as pioneering in this field of research, particularly his works from the early 1960s dealing with variation of hydrological phenomena, their periodicity and the possibility of utilizing the knowledge of that periodicity for formulating long-range forecasts. For instance, in his work on the prognostication of flows [15] Bratránek examined the periodic tendencies in long flow series by applying the methods of moving average and harmonic analysis, and he tried to elucidate the problem of the effect of solar radiation upon the variation of annual precipitates and flows.

He paid particular attention to solar radiation and the effect of solar radiation on hydrological phenomena in his paper [16] published in 1965, where he indicated the complexity of the problem of formulating long-range forecasts. In spite of the fact that for every 11-year solar cycle two hydrological maxima had been recorded, no closer relationship could be detected between the maxima of solar spots and the values of the maxima of precipitation and flows; an indication of a closer relationship was however observed between the values of the maxima of precipitation or flows and the differences between the highest and the lowest values of these maxima.

The examination of the relationship of correlation between the moving statistical characteristics of long flow series and Wolf's average annual numbers characterizing solar activity proved to be rather more promising. These relationships have been studied in more detail by Bratránek [17], Vitha [112], and Souček [102, 103] in long flow series covering a period of more than one hundred years. The closest relationship was found by Souček [102, 103] between the moving coefficients of the average annual flow variations of the Elbe at Děčín and the moving long-term averages of Wolf's numbers (with the mutual coefficient of correlation, r equal to 0.85). That close relationship, schematically visualized in Fig. 39, has prompted some optimism concerning the possibility of long-range forecasting of the river runoff variations based upon the periodic variations of solar activity, and it has also highlighted the relationship between

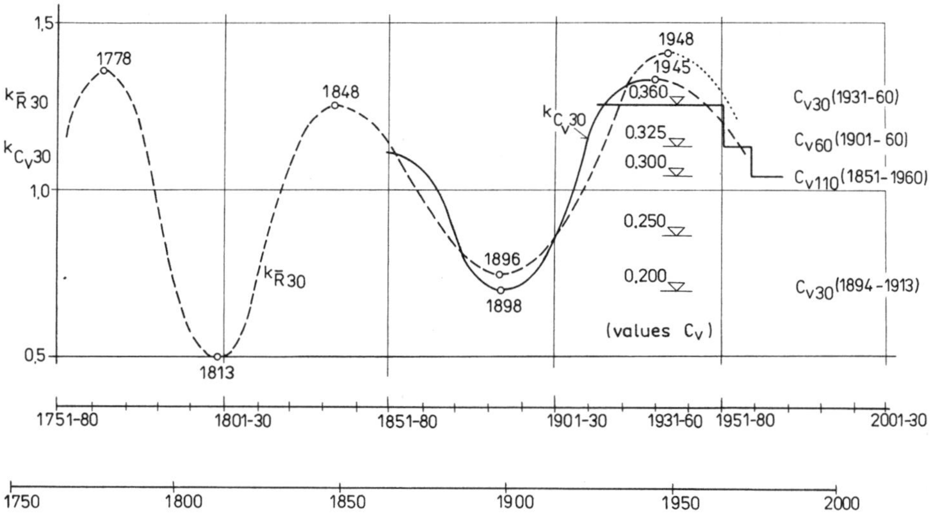

Fig. 39. Long-term fluctuation of solar activity and flows (dashed line – 30-year means of annual Wolf's numbers; continuous line – coefficients of variation $C_{v,30}$ of 30-year samples of the series of average annual flows of the Elbe at Děčín). According to [102, 103].

the sample coefficients of variation and the long-term value (e. g., C_v for the frequently quoted period of the years 1931 to 1960 is relatively high compared with the long-term value).

As far as the eight remaining largest rivers of the world are concerned, the paired coefficient of correlation ranged between 0.35 and 0.79 [102]. These results indicated that with snow- and/or rain-fed rivers the correlation would probably be most pronounced between the variability of the river flowoffs and solar activity, whereas in the other cases, viz. prevailingly glacier-fed flows, the effect of elevated temperatures, balanced runoffs due to a larger number of lakes located in the river-basin etc., the relationship would be much looser or rather insignificant.

Having studied these problems, Bratránek [17] arrived at the following conclusion: the only positive result to arise from the evaluation of the relationship of correlation between the moving characteristics of the flow series and the series of Wolf's numbers is that the fluctuation of the sample coefficients of the flow series variation is a most irregular phenomenon, for which it may be difficult to find any closer correlation to solar activity to use in the formulation of more reliable forecasts.

The latest research, carried out by VÚV (Institute for Water Engineering Research) in Prague (Buřita et al. [23]), dealt systematically with the correla-

TABLE 20. Flow series selected for the examination of bias

Profile	River	Area of river basin (km^2)	Observation period	Parameters of annual flows			
				Q_a (m^3 s^{-1})	C_v	C_s	r (1)
Vilémov	Jizera	146.29	1921–80	4.832	0.234	0.017	0.194
Klášterec n. O.	Divoká Orlice	155.15	1941–80	3.137	0.300	1.083	0.212
Děčín	Elbe	51.103.89	1891–80	311.8	0.308	0.871	0.344
Křivoklát	Berounka	3.422.22	1931–80	32.646	0.445	1.085	0.543
Vestec	Mrlina	460.21	1955–80	1.851	0.645	0.948	0.641

tions between the moving characteristics over a longer period of observation and compared them with the results arrived at twenty years ago; it has shown that with the autocorrelations increasing the moving characteristics acquire complex specific properties. The long-term periodic tendencies and fluctuations of flows can thus in no way be viewed as following some law. From the analyses undertaken it follows that most of the helio-hydrometeorological relationships of correlation vary with different geographical conditions, periods of time and other circumstances.

The results of the research have so far confirmed the considerable difficulties involved in the assessment of the representativeness of the real flow series (particularly those based upon shorter periods of observation), the ascertainment of the corresponding random errors and the estimation of long-range parameters. Under these circumstances, estimation theory approaches the problem in the way described in Part I of this work. With the random errors unknown, the given sample characteristics are viewed as the mean values of the whole population and are cleared of systematic errors.

For our research into bias and systematic errors we have selected five flow series representing the hydrological regimes of rather variable properties (see Table 20). With all the profiles use has been made of the linear regression model for generating a 1 000-year random series of average annual and monthly flows, from which five hundred random samples of various size have been made up. Bias and systematic errors have been examined using the usual moments method.

The unsteadiness of the annual average flows is relatively very low under the climatic conditions of the Czechoslovakia and it can descriptively be approximated to using the interval of variation coefficients according to Table 20. The systematic errors of estimation, C_v, are thus also equal to zero, or they are practically negligible. These properties of the systematic errors of estimation, C_v, are shown schematically in Fig. 40, which represents an extreme model example of the unsteady river Mrlina.

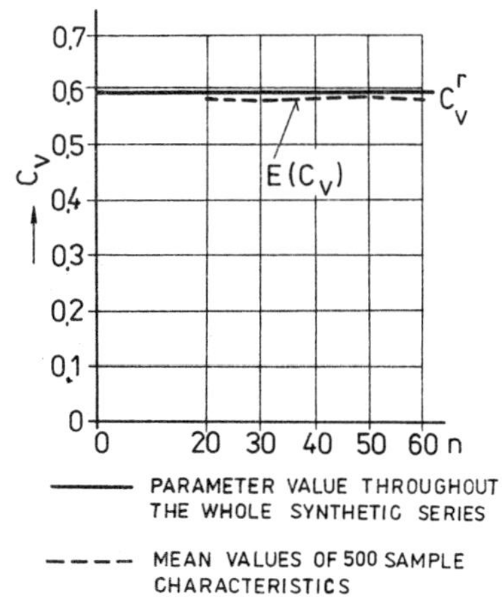

Fig. 40. Systematic errors of coefficients of variation of the series of average annual flows of the river Mrlina.

More attention should be paid to the systematic errors of the coefficients of assymmetry of the average annual flows. As shown in Table 20, their values can vary within a large range – from a near-zero skewness indicating a fairly high degree of normality of distribution, to pronounced skewness of asymmetric distribution, in which the C_s/C_v ratio can even be greater than 3.

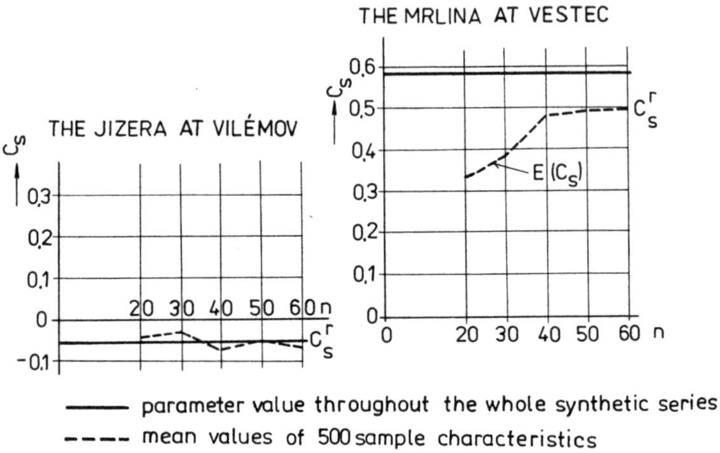

Fig. 41. Systematic errors of coefficients of asymmetry of the series of average annual flows of the rivers Jizera and Mrlina.

Figure 41 presents the curves of systematic errors C_s in two extreme model cases. Relatively minor systematic errors have been found in the annual flow series of the river Jizera, the coefficient of asymmetry of which approximates to zero and the systematic errors are fully negligible. Relatively more pronounced systematic errors C_s have on the other hand been detected in the annual flow series of the rivers Berounka and Mrlina.*)

The research has thus shown that under the climatic and geographical conditions of the Czechoslovakia it will invariably be unnecessary to consider systematic corrections of the sample coefficients of variation of the average annual flows, for their values are relatively low (the systematic errors occurring according to the diagrams worked out with the values of $C_v > 0.60 - 0.70$ and with the sizes of the samples smaller than 20 years). Attention will however have to be given to the systematic errors of the coefficients of assymmetry, which may in no way be negligible and can markedly influence the transgression of the theoretical line marking the boundary of the population.

---

*) From the comprehensive analysis of the characteristics of the average annual flows published in Hydrologické poměry ČSSR (Hydrological Regimes of the Czechoslovak Socialist Republic), vol. III, it follows that the coefficients of asymmetry can reach even higher values than those quoted in the model cases selected.

The apparently simpler problem of the estimation of the parameters of the average annual flows has however some hidden issues, on which it will certainly be useful to focus in further studies. Apart from the problem of representativeness of the given real samples, which has already been analyzed above, there is still the fundamental problem of the estimation of the type of the distribution of the population and its effect upon the parameters to be estimated, which however brings us to the wider problem of the robustness of the estimates. This problem has so far not been satisfactorily researched in hydrology and water engineering.

9.2 Problems of estimation of the autocorrelation function

The bias and the systematic errors of the autocorrelation coefficients of the average annual flows have been examined for five profiles (Table 20). Our methodological approach was analogous to the one applied in the examination of the bias of the sample coefficients of variation and asymmetry.The mutual relationship between the sample autocorrelation coefficients and the autocorrelation coefficients of the population can easily be observed using random sequences modelled for the probability properties stipulated. This advantage of the random sequences facilitates investigation of all the probability properties of the sample autocorrelation coefficients (including their statistical characteristics) in the same way as the probability properties of other sample characteristics.

In modelling the 1 000-year random series the authors paid particular attention to the agreement between the input and the output parameters. Figure 42 (top part) gives a graphical representation of the modelled series for the selected model cases of the river Berounka in the Křivoklát profile and the river Elbe at Děčín. The comparison of the two manifests very good agreement achieved in the modelling of the series. The harmonic shape of the autocorrelation functions is of special interest; it should be taken into account by the design-engineers of storage reservoirs [83].

The probability properties of the first five autocorrelation coefficients have been examined in detail in all the flow series selected. Table 21 presents an example of this examination for the river Berounka at Křivoklát, and the lower part of Fig. 42 shows the curves of the systematic errors.

The model cases examined have shown that the autocorrelation coefficients of the same random series can range within relatively wide limits, so that the values of the random autocorrelation coefficients can be burdened with considerable random errors. Table 21 shows that the extreme values of these coefficients, as well as the range of their variation, $\Delta r(\tau)$, grow rapidly, particularly with the autocorrelation coefficients for greater τ, which invariably point to looser correlative tendencies. In these cases the root-mean-square deviations of the autocorrelation coefficients, particularly their ranges of variation, can even become a higher multiple of the mean value of the set of the sample coefficients itself. It is therefore essential that the significance of the autocorrelation coefficients should be subjected to the respective tests.

The properties of the skewness of the sets of five hundred sample correlation coefficients are of particular interest. Higher values of the coeffcients of asymmetry, $C_s(r(\tau))$, often occur with higher τ's, whereas with lower τ's an indication of a fairly symmetrical distribution can be demonstrated in a number of cases.

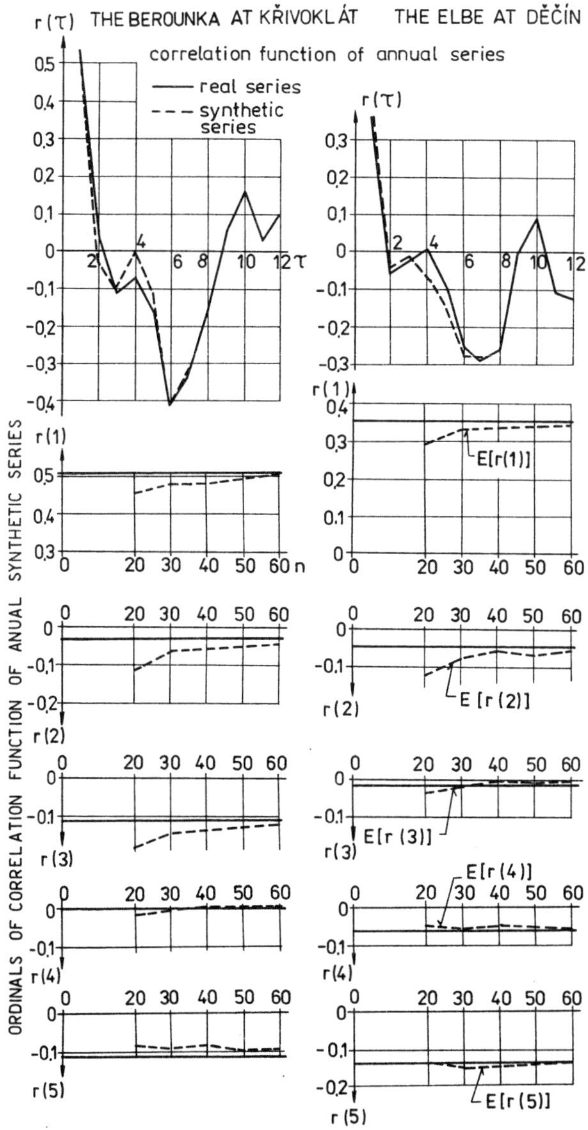

Fig. 42. The autocorrelation function of average annual flows of the river Berounka in the Křivoklát profile and the river Elbe at Děčín, and their systematic errors.

TABLE 21. Properties of the sets of 500 sample autocorrelation coefficients of a series of annual flows in the Křivoklát profile of the river Berounka

Length of samples (years)	$r(\tau)$	$E(r(\tau))$	$\sigma(r(\tau))$	$C_s(r(\tau))$	$r_{max}(\tau)$	$r_{min}(\tau)$	$\Delta r(\tau) = r_{max}(\tau) - r_{min}(\tau)$
20	$r(1)$	0.458	0.161	−0.110	0.786	0.011	0.775
	$r(2)$	−0.117	0.298	0.096	0.593	−0.790	1.383
	$r(3)$	−0.188	0.208	0.294	0.458	−0.630	1.088
	$r(4)$	−0.014	0.277	0.357	0.717	−0.575	1.292
	$r(5)$	−0.032	0.359	0.321	0.791	−0.788	1.579
30	$r(1)$	0.483	0.133	0.039	0.770	0.073	0.697
	$r(2)$	−0.067	0.247	−0.046	0.481	−0.719	1.200
	$r(3)$	−0.148	0.167	0.244	0.399	−0.540	0.939
	$r(4)$	−0.005	0.221	0.540	0.593	−0.460	1.053
	$r(5)$	−0.091	0.295	0.428	0.718	−0.678	1.396
40	$r(1)$	0.486	0.107	0.240	0.739	0.214	0.525
	$r(2)$	−0.061	0.205	−0.061	0.407	−0.560	0.967
	$r(3)$	−0.137	0.136	0.046	0.364	−0.499	0.863
	$r(4)$	0.006	0.181	0.467	0.522	−0.355	0.877
	$r(5)$	−0.080	0.260	0.407	0.597	−0.608	1.205
50	$r(1)$	0.495	0.095	0.283	0.750	0.240	0.510
	$r(2)$	−0.053	0.187	−0.277	0.395	−0.507	0.902
	$r(3)$	−0.131	0.113	−0.116	0.173	−0.453	0.631
	$r(4)$	0.007	0.167	0.510	0.475	−0.297	0.770
	$r(5)$	−0.095	0.233	0.455	0.458	−0.517	0.975
60	$r(1)$	0.503	0.090	0.262	0.742	0.271	0.471
	$r(2)$	−0.043	0.173	−0.192	0.379	−0.498	0.877
	$r(3)$	−0.128	0.108	−0.079	0.189	−0.446	0.635
	$r(4)$	0.006	0.153	0.703	0.447	−0.290	0.737
	$r(5)$	−0.091	0.206	0.564	0.423	−0.418	0.841

In terms of the absolute values the coefficients of asymmetry are however not so high, and it can be concluded that the probability distributions of the individual $r(\tau)$'s will most likely be of a closely similar type.

The properties of the systematic errors of the autocorrelation coefficients (see Fig. 42) very often resemble the properties of the systematic errors of the coefficients of variation and asymmetry.

The dependence of the systematic errors of the ordinates of the correlation function upon the long-term unbiassed value (of the parameter) can be adduced as the first property of this type. It invariably holds that the higher this value, the greater its systematic error. This relationship is particularly conspicuous with the first autocorrelation coefficients. For greater τ's the systematic errors approximate to zero.

The absolute values of the systematic errors of the autocorrelation coefficients are however rather low (as low as hundredths of a unit) in all the cases examined. The relatively greatest systematic errors are manifested by the autocorrelation coefficients of the extraordinarily small-size samples (approximately up to size $n = 20$ years); with an extension of the size of the sample the extent of the systematic errors will become less and invariably approximate to zero. In these cases the autocorrelation coefficients can thus be viewed as being practically unbiassed.

It is obvious that with the autocorrelation functions of the average annual flow series, asymptotically unbiassed estimates can well be proved in a number of cases. So, for instance, the first autocorrelation coefficient, $r(1)$, which is positive in the series selected, has the mean values, $E(r(1))$, one-sidedly deviated below its long-term value, and with the size of the samples increasing they converge towards that value. With the autocorrelation coefficients, $r(\tau)$, the values of which approximate to zero, these relationships are however less pronounced, because the systematic errors are fairly insignificant and they can be both positive or negative.

The properties that have been discussed above prove to be of particular importance as far as mathematical modelling of the annual flow series and the solution of the water- engineering problems, particularly the problems of the design of water storage reservoirs, are concerned. From the results obtained it follows that the empirical sample autocorrelation functions of the average annual flows, with their ordinates viewed as the mean values of the set of sample correlation coefficients, can well be used in the input of the model without the risk of any gross errors.

The disregard of the minor systematic errors is also justified by another fact. As it is well known, modelling of the hydrological series can under no circumstances ensure absolute agreement between the output and the input parameters, and a certain measure of statistically insignificant bias must therefore be reckoned with. Any correction of the input autocorrelation coefficients thus has no practical effect, provided of course the systematic errors are less significant than the random errors.

Great progress has been made in the analysis of the correlation properties of hydrologic series in the past years. Apart from their identification, an assessment of the significance and the extent of their bias is at present fully feasible, and the relationships of correlation can well be included in the domain of mathematical modelling. For further research, one problem has however remained unsolved, viz. the form of the autocorrelation function of the population estimated on the basis of short-term observation. This problem is of course closely linked with what is called the "robustness" of an estimate. Its importance has already been discussed above.

10
Estimation of parameters of average monthly flow series

10.1 Estimation of parameters of probability distribution

The problem of the bias of the characteristics of the average monthly flow series and the estimation of the respective parameters have so far not been satisfactorily elucidated. Research has mostly concentrated upon various methodological procedures of estimating the parameters of the time series of random variables, regardless of their seasonal distribution. These procedures can be applied to the series of culminating or average annual flows. The prevailing interest in the culminating flow series can be attributed to the fact that these series exhibit higher unsteadiness, as well as skewness, resulting in considerable random and systematic errors.

Also, the more complex mathematical models involved in the examination of these relationships have surely contributed to some neglect of the investigation of the properties of the bahaviour of the random and systematic errors in the average monthly flow series. As it is well known, several tens of parameters must be introduced into a linear regression stochastic model of average monthly flows. Apart from the moments of distribution of the probability of flows in the individual calendar moths, it is indispensable that attention should be given to the larger number of correlations between them.

Despite the difficulties mentioned above, research into the extent of bias of the average monthly flows should be regarded as topical, because these series are currently used as the initial data for tackling various important water-engineering problems, as well as a basis on which the mathematical models upgrading the water-engineering computations can be constructed. That is why the check on the representativeness of the average monthly flows and the estimation of their parameters belong to the basic procedures of hydrological data processing.

In our research [78] we investigated the properties of the average monthly flows in a way similar to that applied to the average annual flows (see Chapter 9). For the water measuring profiles (Table 20) we modelled 1 000-year random

TABLE 22. Statistical analysis of the characteristics of five hundred 30-year samples of the 1 000-year series of average monthly flows in the Děčín profile of the river Elbe

Characteristics s		Characteristics of 500 samples					
		$E(s)$	$\sigma(s)$	$C_v\ (s)$	$C_s\ (s)$	max s	min s
Characteristics of the samples of the series of annual flows	Q_a	314.0	16.153	0.051	−0.752	355.5	270.0
	C_v	0.316	0.064	0.202	0.883	0.511	0.210
	C_s	0.595	0.501	0.842	0.708	2.066	−0.516
	$r(1)$	0.330	0.163	0.494	−0.574	0.679	−0.135
	$r(2)$	−0.077	0.213	−2.783	−0.344	0.373	−0.658
	$r(3)$	−0.021	0.163	−7.809	−0.020	0.440	−0.460
	$r(4)$	−0.055	0.231	−4.225	0.128	0.513	−0.537
	$r(5)$	−0.148	0.247	−1.667	0.297	0.518	−0.680
Characteristics of the samples of the series of monthly flows	Q_a	314.0	16.153	0.051	−0.752	355.5	270.0
	C_v	0.680	0.051	0.074	0.166	0.837	0.590
	C_s	1.599	0.248	0.155	0.210	2.363	1.066
	$r(1)$	0.499	0.085	0.130	−0.135	0.661	0.318
	$r(2)$	0.259	0.075	0.289	0.594	0.461	0.106
	$r(3)$	0.083	0.076	0.910	0.365	0.281	−0.092
	$r(4)$	−0.012	0.068	−5.578	0.433	0.176	−0.168
	$r(5)$	−0.081	0.065	−0.804	0.471	0.125	−0.227
Characteristics of the samples of November flows	$Q_{XI,a}$	225.1	17.975	0.080	−0.432	264.1	173.4
	$C_{v,XI}$	0.503	0.065	0.130	0.806	0.689	0.377
	$C_{s,XI}$	0.816	0.428	0.524	1.380	2.656	0.110
	$r_{XI,X}$	0.113	0.218	1.932	0.103	0.701	−0.382
Characteristics of the samples of December flows	$Q_{XII,a}$	268,6	26.804	0.100	−0.238	330.5	174.9
	$C_{v,XII}$	0.590	0.095	0.161	0.117	0.842	0.366
	$C_{s,XII}$	1.017	0.402	0.396	0.091	2.022	0.023
	$r_{XII,XI}$	0.422	0.198	0.496	−0.427	0.817	−0.196
Characteristics of the samples January flows	$Q_{I,a}$	337.8	42.678	0.126	−0.184	458.2	229.4
	$C_{v,I}$	0.643	0.109	0.169	0.513	0.980	0.444
	$C_{s,I}$	1.074	0.428	0.399	0.463	2.555	0.106
	$r_{I,XII}$	0.547	0.186	0.340	−0.425	0.910	−0.017
Characteristics of the samples of February flows	$Q_{II,a}$	420.8	49.145	0.117	−0.007	541.5	319.6
	$C_{v,II}$	0.628	0.089	0.142	0.188	0.862	0.440
	$C_{s,II}$	0.927	0.418	0.449	0.498	2.166	0.101
	$r_{II,I}$	0.443	0.214	0.482	−0.105	0.912	−0.151

Characteristics of the samples of March flows	$Q_{III,a}$	563.9	53.010	0.094	−0.203	700.4	440.3
	$C_{v,III}$	0.526	0.067	0.127	0.261	0.742	0.368
	$C_{s,III}$	0.747	0.304	0.408	0.989	1.838	0.135
	$r_{III,II}$	0.427	0.116	0.273	0.054	0.795	0.084
Characteristics of the samples of April flows	$Q_{IV,a}$	480.9	40.738	0.085	−0.299	561.6	377.0
	$C_{v,IV}$	0.467	0.057	0.123	0.108	0.637	0.342
	$C_{s,IV}$	0.705	0.350	0.496	0.656	1.960	−0.100
	$r_{IV,III}$	0.359	0.191	0.533	−0.917	0.715	−0.381
Characteristics of the samples of May flows	$Q_{V,a}$	339.3	29.997	0.088	−0.178	409.9	272.4
	$C_{v,V}$	0.457	0.048	0.104	−0.123	0.588	0.317
	$C_{s,V}$	0.759	0.351	0.463	0.489	2.061	−0.098
	$r_{V,IV}$	0.401	0.179	0.447	−0.095	0.771	−0.052
Characteristics of the samples of June flows	$Q_{VI,a}$	251.8	25.153	0.100	0.141	324.8	191.7
	$C_{v,VI}$	0.490	0.060	0.122	0.606	0.716	0.336
	$C_{s,VI}$	0.714	0.381	0.534	0.752	1.981	−0.036
	$r_{VI,V}$	0.386	0.168	0.434	−0.175	0.794	−0.053
Characteristics of the samples of July flows	$Q_{VII,a}$	237.4	21.751	0.092	0.103	301.8	179.5
	$C_{v,VII}$	0.557	0.070	0.126	0.195	0.740	0.373
	$C_{s,VII}$	0.798	0.415	0.520	1.248	2.623	−0.078
	$r_{VII,VI}$	0.410	0.175	0.428	−0.386	0.783	−0.093
Characteristics of the samples of August flows	$Q_{VIII,a}$	207.6	23.615	0.114	−0.444	267.1	138.7
	$C_{v,VIII}$	0.589	0.073	0.123	0.708	0.844	0.451
	$C_{s,VIII}$	0.942	0.347	0.369	0.450	2.283	0.138
	$r_{VIII,VII}$	0.409	0.153	0.375	0.049	0.719	0.073
Characteristics of the samples of September flows	$Q_{IX,a}$	211.8	22.527	0.106	0.343	278.9	151.8
	$C_{v,IX}$	0.570	0.070	0.123	−0.156	0.750	0.308
	$C_{s,IX}$	0.835	0.393	0.470	−0.003	1.855	0.004
	$r_{IX,VIII}$	0.433	0.145	0.335	−0.069	0.785	0.007
Characteristics of the samples of October flows	$Q_{X,a}$	223.5	20.185	0.090	−0.280	273.2	171.9
	$C_{v,X}$	0.513	0.070	0.136	0.059	0.728	0.351
	$C_{s,X}$	0.865	0.362	0.418	0.193	1.840	−0.045
	$r_{X,IX}$	0.378	0.175	0.463	−0.478	0.767	−0.161

series of average monthly flows, from which sets of 500 random samples were made up with lengths of 20, 30, 40, 50 and 60 years. This interval satisfactorily covers the length of observation that is currently applied in practice. The samples were generated in three ways: with their starting years randomly selected and the chronological arrangement of all its elements retained; with the annual flows absolutely random and the distribution of the monthly flows in each year retained; and in a moving way, viz. according to a certain rule of selecting the beginning years of the samples. We regarded the first way of generating samples, with the beginnings randomly selected, as fundamental, the other two ways having been tested on a single model case only.

After the selection of the methodological approach to the examination of the properties of the bias and the estimates, then logicaslly followed the method and the scope of the computations of the moment characteristics and their statistical processing. As compared with the series of average annual flows it becomes necessary in this case to tax the computer with a very much greater amount of computations, for with samples of the monthly flow series not only the characteristics of the chronological series and the characteristics of the flows in the individual calendar months must be considered, but also their correlations. In our case, we computed the first three moment characteristics $(Q_{m,a}, C_{v,m}, C_{s,m})$ and the coefficients of correlation $r_{m,m-1}$ between the flows in the given month and those in the preceding month. In a way similar to the one applied in the case of the chronological arrangement of the elements of the samples, we processed statistically the 500-element sets of characteristics and compared them with the corresponding parameters ascertained throughout the whole 1 000-year random series.

Table 22 is an example of the statistical analysis of the sets of five hundred 30-year characteristics of a modelled series of average monthly flows in the Děčín profile on the river Elbe. The analysis was then repeated for each of the lengths of the samples, viz. for 20, 30, 40, 50 and 60-year periods.

The most significant results, particularly the dependence of the systematic errors upon the length of the sample, was given a graphical form, which greatly facilitated checks on the measure of bias of the individual characteristics sought. The value of the output parameter of the modelled 1 000-year series was considered to be the unbiassed estimate of the respective parameter. The expected values of the sets of 500 sample characteristics were then taken to be biassed characteristics.

The results of the investigation of the properties of bias and systematic errors in five model cases can be summed up as follows:

1. The systematic errors of the coefficients of variation in the individual calendar months depend above all upon the unsteadiness of the flow regime. The dependence upon the length of the sample is rather less marked and it becomes more pronounced with unsteadiness of higher degrees. The characteristic exam-

ple in Fig. 43 (the river Berounka in the Křivoklát profile) shows that the systematic errors of the coefficients of variation are practically negligible up to about values of $C_v = 0.60$ to 0.70. More pronounced systematic errors appear

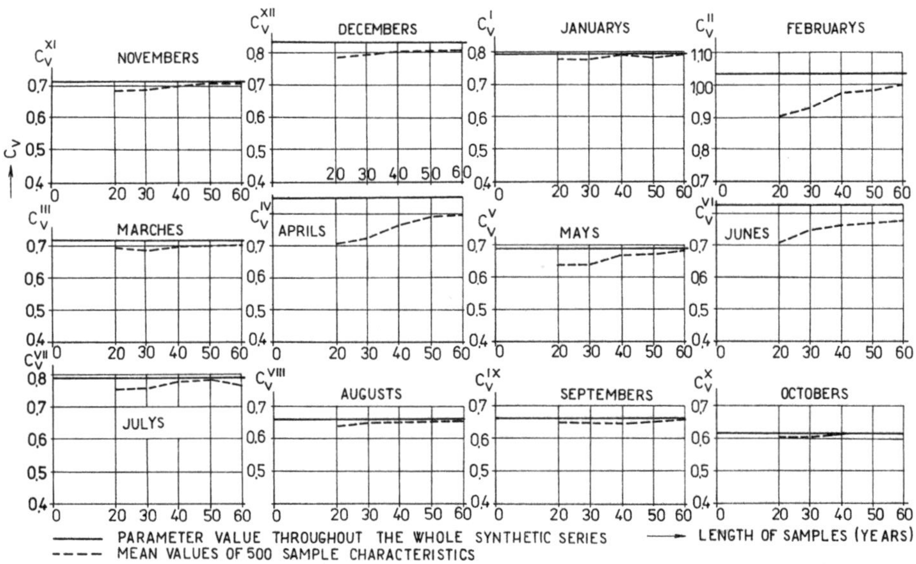

Fig. 43. Systematic errors of coefficients of variation of the average flows in the calendar months in the Křivoklát profile on the river Berounka.

with higher degrees of unsteadiness of the flow regime (viz. the river Berounka particularly in February, April and June).

These results are in agreement with the general knowledge of the behaviour of systematic errors C_v, and they are also confirmed by the properties of the series of the average annual flows. The mutual relationship between the mean values of the sets of sample characteristics and the parameters is of the usual character and it practically confirms the asymptotic behaviour of the estimates (the systematic errors growing less with the length of the sample, n, increasing).

2. The investigation of the systematic errors of the coefficients of variation of the chronological sequences of average monthly flows did not provide any new information. In spite of the fact that the unsteadiness of these series is higher than the unsteadiness of the average annual flow series, the systematic errors are almost negligible. This property can be accounted for by the fact that with the length of the samples (viz. the number of years) constant, the number of their elements is twelve times higher, which of course results in smaller systematic errors.

3. The systematic errors of the coefficients of asymmetry of the average flows in the individual calendar months depend, like systematic errors C_v, upon the value of their parameter, viz. upon the coefficient of asymmetry of the given

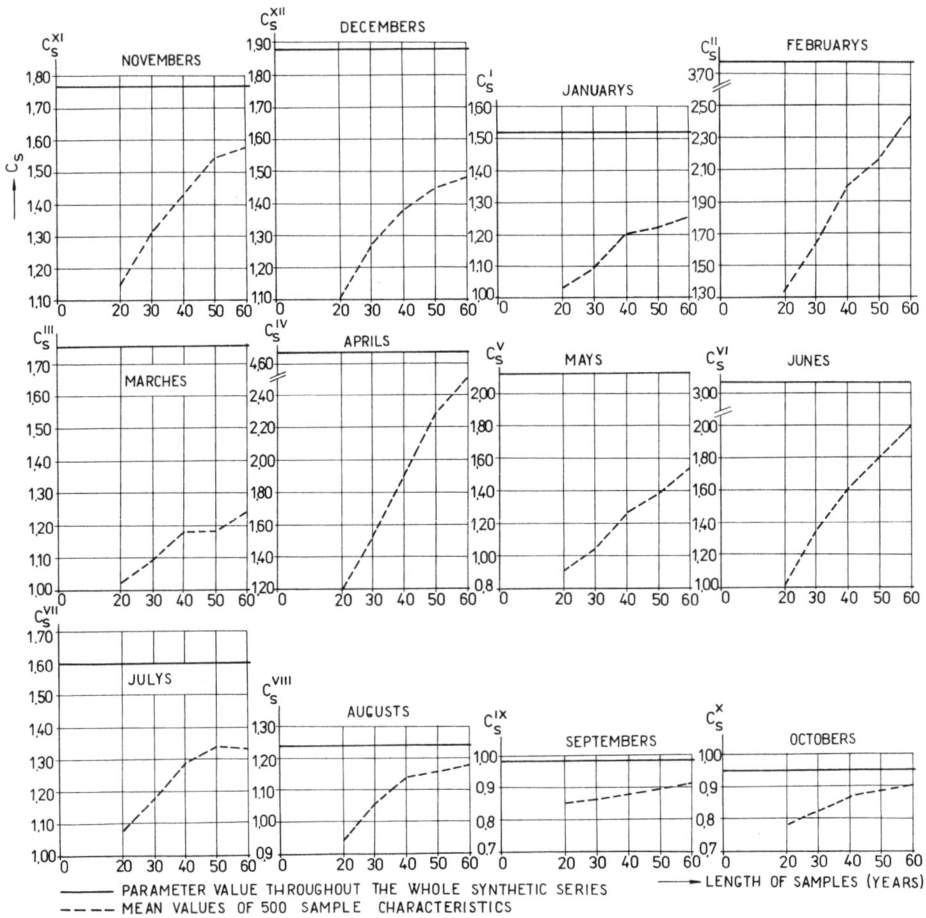

Fig. 44. Systematic errors of coefficients of asymmetry of the average flows in the calendar months in the Křivoklát profile on the river Berounka.

series. Unlike systematic errors C_v, they were however proven in all the five modelled cases. Also their dependence upon the length of the sample is clearly defined. The example in Fig. 44 (the river Berounka in the Křivoklát profile) confirms the asymptotic behaviour of systematic errors. It is obvious that they ought to be paid attention to, particularly with the shorter samples.

From among the model cases, the relatively least systematic errors C_s could be found in the monthly flows of the rivers Jizera and Orlice, where the skewness of the modelled series was not so high (the values of the C_s/C_v ratio for the individual months equalling approximately 1–2). And the flows of the river Mrlina have the greatest systematic errors, where the C_s/C_v ratio invariably exceeds 2. In these cases systematic errors can reach several tens percent, or a value equal to the mean value of the set of sample C_s's.

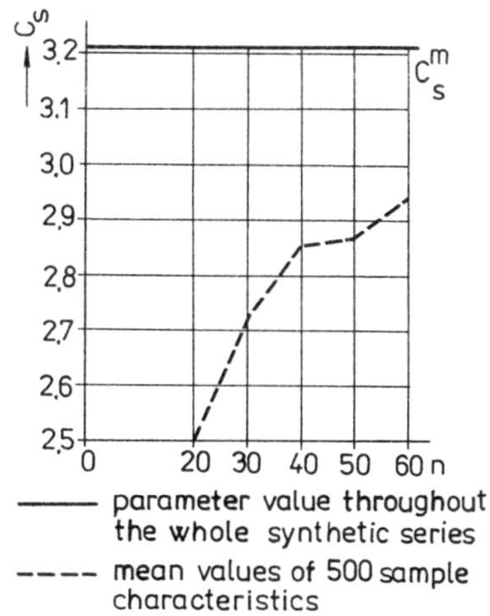

Fig. 45. Systematic errors of coefficients of asymmetry of average monthly flows of the river Mrlina.

4. In the average monthly flow series, systematic errors should even be checked in case they are chronologically arranged. The example in Fig. 45 (the river Mrlina in the Vestec profile) proves that with the skewness of the distribution more pronounced, systematic errors are in no way negligible.

10.2 Problems of estimation of the autocorrelation function

The probability properties of the sample autocorrelation coefficients of the average monthly flow series were derived in a way analogous to that used in the case of the average annual flows. With the individual autocorrelation coefficients we concentrated on both the curves of bias and the curves of systematic errors, as well as the remaining statistical characteristics.

The curve of the autocorrelation function of the average monthly flows of the river Mrlina in Fig. 46 shows that the systematic errors are very small with all the ordinates. This is mainly accounted for by the fact that the samples of average monthly flow series invariably have a sufficiently large number of elements (compared with the same samples of annual flows the number of elements, thus also the number of correlated pairs, is twelve times higher). The ordinates of the autocorrelation function can thus, under these circumstances, be regarded as approximately unbiassed.

The properties of the moment characteristics of the autocorrelation coefficients are even more interesting. The example of the river Berounka in the

Křivoklát profile (Table 23) shows that the sample autocorrelation coefficients of the average monthly flow series can be burdened, like the autocorrelation coefficients of the average annual flow series, with considerable random errors,

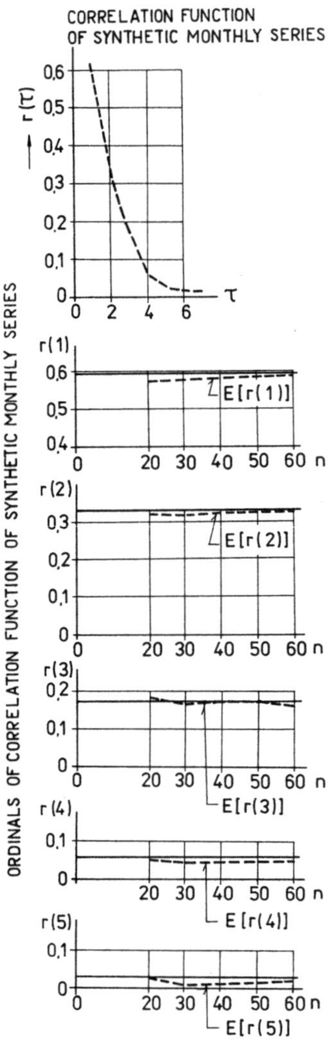

Fig 46. The autocorrelation function of average monthly flows of the river Mrlina, and its systematic errors.

particularly with the looser correlation tendencies. This is well proved by the standard deviations $\sigma(r(\tau))$ and the range of variation $\Delta r(\tau)$. It follows that the problem of estimating the form of the autocorrelation function from a single, particularly shorter, observation is extraordinarily complex and deserves receive attention in future research.

It is certainly worthy of mention that both the standard deviations and the variation range of the autocorrelation coefficients are lower with the monthly

TABLE 23. Properties of the sets of 500 sample autocorrelation coefficients of a series of monthly flows in the Křivoklát profile of the river Berounka

Length of sample (years)	$r(\tau)$	$E(r(\tau))$	$\sigma(r(\tau))$	$C_s(r(\tau))$	$r_{max}(\tau)$	$r_{min}(\tau)$	$\Delta r(\tau) = r_{max}(\tau) - r_{min}(\tau)$
20	$r(1)$	0.498	0.095	0.061	0.755	0.256	0.499
	$r(2)$	0.241	0.082	0.385	0.513	0.053	0.460
	$r(3)$	0.094	0.070	0.500	0.289	−0.047	0.336
	$r(4)$	−0.016	0.074	0.364	0.188	−0.187	0.375
	$r(5)$	−0.079	0.079	0.346	0.172	−0.271	0.443
30	$r(1)$	0.493	0.077	0.139	0.723	0.323	0.400
	$r(2)$	0.242	0.062	0.411	0.467	0.103	0.364
	$r(3)$	0.100	0.058	0.444	0.285	−0.029	0.314
	$r(4)$	−0.003	0.065	0.412	0.167	−0.142	0.309
	$r(5)$	−0.068	0.068	0.374	0.143	−0.208	0.351
40	$r(1)$	0.490	0.072	0.218	0.691	0.356	0.335
	$r(2)$	0.240	0.057	0.678	0.439	0.118	0.321
	$r(3)$	0.098	0.050	0.692	0.268	0.009	0.259
	$r(4)$	−0.006	0.057	0.330	0.151	−0.151	0.302
	$r(5)$	−0.063	0.062	0.401	0.115	−0.217	0.332
50	$r(1)$	0.492	0.065	0.328	0.660	0.379	0.281
	$r(2)$	0.247	0.054	0.749	0.419	0.133	0.286
	$r(3)$	0.107	0.047	0.827	0.250	0.028	0.222
	$r(4)$	−0.000	0.051	0.125	0.122	−0.113	0.235
	$r(5)$	−0.058	0.052	0.262	0.113	−0.181	0.294
60	$r(1)$	0.484	0.058	0.550	0.834	0.395	0.239
	$r(2)$	0.243	0.045	0.982	0.397	0.149	0.248
	$r(3)$	0.104	0.040	0.885	0.237	0.028	0.209
	$r(4)$	−0.003	0.046	0.242	0.108	−0.124	0.232
	$r(5)$	−0.060	0.049	0.506	0.076	−0.168	0.244

flow series. This property can be attributed to the genetic tendencies of the hydrological regimes under Czechoslovak conditions, for which closer autocorrelation relationships can often be found with the seasonal distribution of the runoff during the year. In spite of the fact that the distribution of the runoffs in the individual year may vary, the tendency following from seasonal variations can be characterized as analogous.

10.3 Estimation of the coefficients of correlation between the average flow series in calendar months

The relations of correlations between the average flow series in the individual calendar months must be considered in the design of their mathematical models. Their bias and their representativeness are therefore of equal importance as the bias and the representativeness of the other characteristics.

As is well known, the coefficients of correlation between all the combinations of the series of monthly flows can formally be arranged into a correlation matrix, the elements of which can express correlations between the flow series within one hydrological year, or within several preceding years. Their properties are dealt with in detail in one of our studies [84], where we showed that the relatively closest correlative relationships invariably occur between the neighbouring monthly flows. It was these coefficients that the investigation of bias was therefore directed at.

In dealing with their properties we proceeded similarly as in other cases, but the investigation of systematic errors did not provide us with any interesting information, because in all the models under examination systematic errors proved to be very small and practically negligible.

The mutual relationship between the biassed and unbiassed estimates of the coefficients of correlation between the neighbouring monthly flows was also

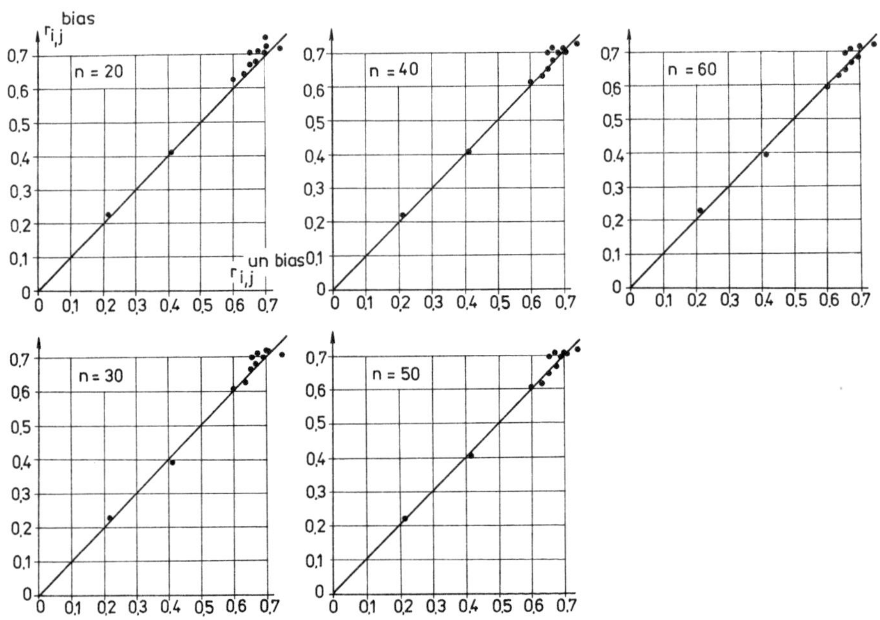

Fig. 47. Relationship between the biassed and the unbiassed estimates of the coefficients of correlation between the neighbouring monthly flows in the Vestec profile on the river Mrlina.

observed for each water-measuring station and each length of the samples selected. We proceeded such that in one correlation field we always plotted the relationships between twelve pairs of coefficients of correlation of the neighbouring monthly flows in the cycle of one hydrological year. Figure 47 shows an example of these relationships for the unsteady river Mrlina. On the horizontal axis are plotted the long-term unbiassed values of the coefficients of correlation of the neighbouring monthly flows, the vertical axis gives the biassed mean values of the sets of 500 sample coefficients of correlation. The differences between these mean values and the corresponding ordinates of the straight lines passing through the origin with their gradient equal to unity are systematic errors. From the graphical representation it can be seen that these errors are negligible as far as the range of the sample lengths $n = 20$ to 60 years is concerned.

The results of the research show that the empirical sample coefficients of correlation between neighbouring monthly flows can approximately be regarded as satisfactory estimates of the long-term values of these coefficients, which can be included in the input of the respective mathematical models.

The results of the research fully justify our assumption that the systematic errors of the other elements of the correlation matrix, which mostly express looser relationships, will be even less, which is why we ignored them.

10.4 Problems of generating random samples from flow series

The process of generating random samples from real or modelled flow series is dealt with in statistical and water engineering literature particularly in the context of the discussion of non-stationarity or non-ergodicity of random processes, or sample surveying and estimation. The importance of examining various random samples is quite obvious: it is a logical inference from the fact that samples have variable probability properties and that they can thus variously affect the solution of hydrological and water-engineering problems.

The possibility of generating random samples from real flow series depends upon the length of the observation. If only a short observation series (e. g. several years only) is available, generating shorter samples is of no practical use, because the series itself has the character of a single sample from the characteristics of which we try to infer long-term unibassed parameters.

Substantially greater possibilities of investigating the properties of samples are offered by the random series that can be modelled in arbitrary lengths. In Part I of this book we have shown that all the relationships of the samples to the whole series approximating to the population (theoretically of infinite length) can reliably be defined on the basis of a satisfactorily large set of samples.

The random samples themselves can be generated in several ways according to pre-formulated rules, which may, to a certain extent, affect some of the probability properties of these samples, and their bias. The methods of generating random samples therefore need to be considered.

In our research we were concerned with three methods of generating random samples of modelled monthly flow series:

1. samples were generated so that their origins (viz. years) were chosen at random, the sequence of their elements corresponding to the order of these elements in the modelled series;
2. samples were generated as absolute random sequences of the annual flows, the distribution of the monthly flows in each year being adhered to in accordance with the original modelled series (viz. each year retaining its own fragment);
3. samples were generated in a moving way, with the origins of the samples chosen according to a given rule.

The first two methods of generating random samples were compared and the probability properties were assessed in the series of average annual flows.

Since the values of the moment characteristics ($\bar{x}$, C_v, C_s) are independent of the order of the elements of the sequence, the expected values of the set of characteristics from which systematic errors are derived do not, of course, vary with a change in the generation of the random samples. The checking computations that were carried out showed full agreement between the expected values of the characteristics of the set arrived at using the two variants of random sample generation. (Next-to-zero deviations can occur if the samples generated using the two variants described do not correspond to each other, that is, if they are asynchronous).

Similar properties are exhibited by the characteristics of the samples of the monthly flow series (i. e. chronologically arranged series) and the series of the flows in the individual calendar months. In this case also the order of their elements does not affect the values of the characteristics of the distribution either.

The correlation properties of the annual and monthly flow series with the order of their elements changed are of particular interest. The autocorrelation relations change most markedly if the samples of annual flow series are generated according to the law of abolute randomness. If the linear regression model closely matches the relations of correlation of the given real series, then the second variant of generating samples will invariably yield statistically insignificant (next-to-zero) mean values of the set of autocorrelation coefficients, which could well be expected.

The autocorrelation coefficients of shorter samples of an absolutely random sequence can of course range within relatively wide limits, from positive to negative values. For their 500-element sets we also computed, apart from the expected values required for the ascertainment of the systematic errors, the other statistical characteristics, including the maximum and the minimum values of the set.

TABLE 24. Comparison of the characteristics of a series of average annual flows in the Vestec profile of the river Mrlina with the parameters of the random series

Series	Characteristics (parameters)									
	$\bar{Q}_a$ ($m^3 s^{-1}$)	C_v	C_s	$r(1)$	$r(2)$	$r(3)$	$r(4)$	$r(5)$	$r(6)$	$r(7)$
Real (in the 1955–1980 period)	1.851	0.645	0.948	0.641	0.243	−0.064	−0.193	−0.380	−0.604	−0.565
Random 1 000–year series	1.820	0.595	0.582	0.618	0.238	−0.051	−0.160	−0.335	−0.554	−0.527

TABLE 25. Correlation properties of a set of 500 samples formed as absolute random sequences

Length of samples n (years)	Charac-teristics	Characteristics of 500 samples				
		$E(r(\tau))$	$\sigma(r(\tau))$	$C_s(r(\tau))$	$r_{max}(\tau)$	$r_{min}(\tau)$
20	$r(1)$	−0.040	0.216	0.066	0.621	−0.645
	$r(2)$	−0.039	0.223	0.111	0.655	−0.615
	$r(3)$	−0.039	0.236	0.126	0.590	−0.617
	$r(4)$	−0.071	0.239	0.138	0.684	−0.732
	$r(5)$	−0.057	0.253	0.270	0.642	−0.685
30	$r(1)$	−0.051	0.188	0.070	0.567	−0.594
	$r(2)$	−0.024	0.186	−0.134	0.493	−0.661
	$r(3)$	−0.042	0.192	−0.129	0.522	−0.628
	$r(4)$	−0.029	0.195	0.120	0.487	−0.544
	$r(5)$	−0.030	0.194	0.020	0.507	−0.651
40	$r(1)$	−0.025	0.155	−0.061	0.548	−0.484
	$r(2)$	−0.016	0.156	−0.036	0.410	0.393
	$r(3)$	−0.024	0.160	0.045	0.450	−0.437
	$r(4)$	−0.032	0.158	−0.003	0.454	−0.436
	$r(5)$	−0.028	0.160	−0.037	0.437	−0.532
50	$r(1)$	−0.011	0.135	−0.054	0.364	−0.352
	$r(2)$	−0.025	0.145	0.129	0.522	−0.441
	$r(3)$	−0.020	0.143	−0.017	0.405	−0.468
	$r(4)$	−0.022	0.149	0.014	0.449	−0.346
	$r(5)$	−0.019	0.144	0.061	0.393	−0.368
60	$r(1)$	−0.034	0.135	0.077	0.309	−0.437
	$r(2)$	−0.020	0.128	0.093	0.420	−0.366
	$r(3)$	−0.020	0.131	0.083	0.439	−0.385
	$r(4)$	−0.030	0.140	0.102	0.444	−0.419
	$r(5)$	−0.017	0.132	0.103	0.341	−0.439

An example of the examination of these relationships can be found in Tables 24 and 25. Table 24 compares the characteristics of the given real series of the average annual flows in the Vestec profile of the river Mrlina with those of the random 1 000-year series. The parameters of the modelled series show good agreement with the given characteristics. If we disregard the order of the elements of this sequence and generate samples of it following the law of absolute randomness, the coefficients of autocorrelation and other characteristics can also be found for each sample and sets of samples (viz. always for the length of the sample chosen). The results of these investigations are presented in Table 25. The expected values $E(r(\tau))$ and extreme values $r_{max}(\tau)$ and $r_{min}(\tau)$ will understandably be of the greatest interest. In spite of the fact that the expected values invariably approximate to zero, the individual values of $r(\tau)$ range within relatively wide limits, which narrow with the length of the samples increasing (the standard deviations also grow with n declining).

The results presented show the importance of the solution of the reverse task, viz. estimating the long-term values of the autocorrelation coefficients from the given sample. From the example it follows that a single sample value need in no way correspond to the parameter, and that is why statistical tests of significance (see also Section 9.2) are of such great importance.

The autocorrelation properties of the samples of random monthly flow series undergo very little variation if the order of the years is changed and the expected values of the set of sample autocorrelation coefficients approximate to the expected values of the samples of the regression sequence. This result of the solution can be explained by the fact that the change of the order of the fragments (years) causes a disturbance of the original autocorrelations of the monthly flows on the boundaries of the fragments only. Of course, the measure of the bias of the autocorrelation function depends upon argument τ of coefficients $r(\tau)$; for example, in the computation of $r(1)$ only one out of the twelve pairs of elements undergoes a change on the fragment boundaries.

And similarly, as far as the coefficients of correlation between the neighbouring monthly flows are concerned, if the first and the second variant of the generation of random samples are compared, only the coefficients of correlation between the monthly flows undergo variation on the boundaries of hydrological years.[*)]

Research into the properties of the moving random samples, which we compared with the properties of the samples generated in the two ways mentioned above, did not yield any surprising results. The values of statistical characteristics, including correlation coefficients of the samples of annual and monthly series, approximated to the characteristics of the samples with randomly chosen origins.[**)]

From the analyses carried out it follows that in generating random samples we should primarily concentrate on the autocorrelation properties of the samples, which sounds quite logical. As far as the average monthly flow series are concerned, the correlations are impaired on the boundaries between the individual years provided only the order of the fragments varies but the order of the flows within the fragments remains constant.

---

*) In the fragment method of Svanidze [107] the correlations between the other series of monthly flows also suffer disturbance, for each real fragment gets linearly transformed at a different ratio of yearly flows.

**) The problems of the moving characteristics themselves are of course mathematically rather complex if the sequences of these characteristics are compared with the original time series. The moving characteristics can on the one hand provide information not intrinsic to the original series, on the other hand, some of the properties of the original series can be smoothed out.

11
Automated parameter estimation and computer-aided modelling of random hydrological series

11.1 Automated computer-aided estimation of parameters

The estimation of unbiassed parameters using the moments method is greatly facilitated by the diagrams worked out for the various types and parameters of distribution. However, for bulk processing of the estimates, and particularly for automatic calculations aided by computers, diagrams are in no way so suitable. In the past few years, an intensive search has been made for such analytic expressions that would be easily programmable and would express simply and explicitly the relationships between the given biassed characteristics and the respective unbiassed parameters.

As with the plotting of diagrams, attempts are being made to derive these expressions for the different types of parameters of distribution. In Section 4.2 we mentioned that in this respect advantage was often taken of modelling methods and of the random sequences in which the relationships sought can relatively easily be expressed with the help of computer technology. Whenever analytic expressions are derived, zero value of the first autocorrelation coefficient, viz. $r(1) = 0$, is invariably assumed in an oversimplified way.

For Pearson's IIIrd type distribution the analytic relationships of parameter estimation were derived by Rozhdestvenskiĭ [98]. He found the following relationship for the unbiassed estimator of the coefficient of variation:

$$C_V^* = \frac{a_2}{n} + \left(a_3 + \frac{a_4}{n}\right) C_V + \left(a_5 + \frac{a_6}{n}\right) C_V^2, \tag{11.1}$$

TABLE 26. Values of the coefficients in equation (11.1)

C_s/C_v	a_2	a_3	a_4	a_5	a_6
2	0.19	0.99	−0.88	0.01	1.54
3	0.69	0.98	−4.34	0.01	6.78
4	1.36	1.02	−9.68	−0.05	15.55

where C_v^* stands for the unbiassed estimate of the coefficient of variation, C_v for the sample coefficient of variation, and n for the length of the sample (i. e. the number of the terms of the series).

Coefficients a_2, a_3, a_4, a_5 and a_6 can be read from Table 26.

For the unbiassed estimator of the coefficient of asymmetry Rozhdestvenskiĭ derived the following relationship:

$$C_s^* = \left(0.03 + \frac{2}{n}\right) + \left(0.92 - \frac{5.09}{n}\right) C_s + \left(0.03 + \frac{8.10}{n}\right) C_s^2, \qquad (11.2)$$

where C_s^* stands for the unbiassed estimate of the coefficient of asymmetry, and C_s for the sample coefficient of asymmetry.

A certain disagreement between the values of C_v^* according to equation (11.1) and C_s^* according to equation (11.2), and the original values in the diagrams can be accounted for by the fact that with the computation of the unbiassed estimate of C_v^* according to equation (11.1) the estimation error grows with the values of ratio C_s/C_v and autocorrelation coefficient $r(1)$ increasing. As far as their higher values are concerned, diagrams therefore prove to be more suitable. Rozhdestvenskiĭ does not consider errors in the computation of C_s^* according to equation (11.2) to be significant (unless they exceed the value of 0.1).

The estimation of the coefficient of asymmetry was also dealt with by Bobée and Robitaille [12]. For Pearson's type III distribution they derived the following equation,

$$C_s^* = C_s\left[\left(1 + \frac{6.51}{n} + \frac{20.2}{n^2}\right) + \left(\frac{1.48}{n} + \frac{6.77}{n^2}\right) C_s^2\right], \qquad (11.3)$$

and for the triparametric log-normal distribution equation

$$C_s^* = C_s\left[\left(1.01 + \frac{7.01}{n} + \frac{14.66}{n^2}\right) + \left(\frac{1.69}{n} + \frac{74.66}{n^2}\right) C_s^3\right] \qquad (11.4)$$

where C_s are again the given sample values.

The equation holds for $20 \leqq n \leqq 90$ and $0.25 \leqq C_s^* \leqq 5$.

In the same paper Bobée and Robitaille quote an analogous expression for the estimation of C_s^* with Weibull's distribution, viz.

$$C_s^* = C_s\left[\left(1.01 + \frac{5.05}{n} + \frac{20.13}{n^2}\right) + \left(\frac{0.69}{n} + \frac{27.15}{n^2}\right) C_s^3\right] \qquad (11.5)$$

which holds for the same domains of n and C_s^*.

The mutual relation between C_s^* estimation for Pearson's type III distribution according to Rozhdestvenskiĭ and according to Bobée was studied in detail by Kašpárek et al. [44]. Applying the method of comparative analysis Kašpárek pointed out that Bobée calculated the coefficient of asymmetry using the expression

$$C_{s(1)} = \frac{\sum_{i=1}^{n} (x_i - \bar{x})^3}{ns^3}, \tag{11.6}$$

where

$$s = \left[\frac{\sum_{i=1}^{n} (x_i - \bar{x})^2}{n} \right]^{1/2}, \tag{11.7}$$

whereas Rozhdestvenskiĭ bases his calculations upon the expression

$$C_{s(2)} = \frac{\sum_{i=1}^{n} (x_i - \bar{x})^3 \, n}{(n - 1)(n - 2)\, s^3} \tag{11.8}$$

where

$$s = \left[\frac{\sum_{i=1}^{n} (x_i - \bar{x})^2}{n - 1} \right]^{1/2}. \tag{11.9}$$

A most valuable result of the analysis [44] is the finding that for the number of the terms of the series $n > 30$ the results are nearly identical, and even in the region of $n < 30$ they do not differ by more than 10 percent. In the range considered, Bobée's equation is thus in better agreement with Rozhdestvenskiĭ's results than his own analytic expression. From this point of view it proves more convenient to use equation (11.3) than (11.2).

We oriented our research towards the derivation of analytic expressions for the estimation of parameters of triparametric log-normal distribution, which is practically applied to the solution of a number of important problems. We approached the relationship between the unbiassed estimates of parameters and the characteristics with the help of 10 000-element random sequences modelled with pre-determined probability properties. In order to obtain the most fitting shape of curves of type $C_v^* = f(C_v)$ and $C_s^* = f(C_s)$ use was made of the least squares method.

For the estimation of the coefficients of variation, C_v, we got the following resultant equations:

for $C_s = C_v$:

$$C_v^* = C_v \quad \text{(unbiassed expected values of characteristics)}, \tag{11.10}$$

for $C_s = 2C_v$:

$$n = 20: \quad C_v^* = 1.07C_v - 0.035\,, \tag{11.11}$$

$$n = 60: \quad C_v^* = 1.03C_v - 0.015\,, \tag{11.12}$$

for $C_s = 3C_v$:

$$n = 20: C_v^* = -0.002\,38 + 1.028C_v + 0.013\,4C_v^2 + 0.088\,9C_v^3\,, \tag{11.13}$$

$$n = 60: C_v^* = 0.000\,63 + 0.968\,8C_v + 0.076\,8C_v^2 + 0.005\,9C_v^3\,, \tag{11.14}$$

for $C_s = 4C_v$:

$$n = 20: C_v^* = -0.001\,98 + 1.02C_v + 0.034\,9C_v^2 + 0.118\,5C_v^3\,, \tag{11.15}$$

$$n = 60: C_v^* = 0.000\,24 + 0.993\,2C_v + 0.013\,4C_v^2 + 0.071\,1C_v^3\,. \tag{11.16}$$

In view of the estimation accuracy achievable, the C_v^* values for the intermediate values of n can be approximated by interpolation; for $n > 60$ use can be made without any more serious inaccuracies being incurred of linear extrapolation up to the extreme value of $C_v^* = C_v$, where the systematic errors are zero.

Equations (11.10) to (11.16) were derived from random sequences modelled for the input coefficients of variation within the following limits: $C_v = 0.25 - 1.75$. These values can roughly be regarded as the limiting conditions of C_v estimation.

For the estimation of the coefficient of asymmetry, C_s^*, of the triparametric log-normal distribution we formulated the following equations:

for $n = 20$:

$$C_s^* = -0.000\,592 + 1.003C_s + 0.259C_s^2 + 0.373\,C_s^3\,, \tag{11.17}$$

for $n = 30$:

$$C_s^* = -0.002\,14 + 0.714C_s + 0.586C_s^2 + 0.06C_s^3\,, \tag{11.18}$$

for $n = 60$:

$$C_s^* = 0.011\,09 + 0.708C_s + 0.386C_s^2 + 0.002C_s^3\,. \tag{11.19}$$

Like equation (11.4) derived by Bobée and Robitaille, equations (11.17), (11.18), (11.19) do not depend upon the ratio C_s/C_v.

In this case, too, the conditions limiting the estimation of C_s follow from the range of the input parameters of the models applied. For the shortest length of the samples, $n = 20$, estimate $C_s^* = 4.81$ pertains to the highest sample value, viz. $C_s = 1.80$; for the maximum length of the samples, $n = 60$, the estimate of the highest sample value, $C_s = 2.55$, equals $C_s^* = 4.34$. These limits of estimating C_s^* correspond approximately to $C_s^* = 5$ quoted by Bobée and Robitaille for the validity of equation (11.4).

With the extreme values of C_s in the real series of the average monthly flows, mechanical estimates of C_s^* can lead to unjustifiable values. In these cases an individual analysis of the estimate is advisable, taking due account of the genetic factors biassing the extreme values of C_s, the development of skewness in the neighbouring months, and the effect of extraordinary floods on the values of the moment characteristics; also the regional factors must be considered, viz. the development of skewness recorded by the flow measuring stations in the surroundings etc.

An interesting result came from the comparison of the estimates of C_s^* according to equation (11.4) and equations (11.17) to (11.19). For the lower values of C_s the results of the two methods of solution proved nearly identical; for the higher values of C_s the results of our research contain slightly greater systematic errors.

11.2 The linear regression stochastic model and its modifications

The linear regression stochastic model has been described in detail several times in Czechoslovak and foreign statistical literature. It was introduced into Czechoslovak hydrological literature by Kos [55], who dealt with the theoretical foundations of the model and its application to hydrology. We are therefore not going to derive the model again, and will focus only on the properties that can contribute to the achievement of satisfactory agreement between the input and the output parameters. The following Section 11.3, then deals with the possibilities of modelling random flow series with respect to the bias of the characteristics of real series.

The linear regression stochastic model has been practically tested in detail by various Czechoslovak flow recording stations in the past fifteen years. It turned out that the parameters of the modelled series of the annual and monthly flows agreed fairly well (viz. within the limits of admissible random errors) with the given input parameters and that the modelled series can thus be used as a more adequate instrument for the solution of hydrological problems. Random series are in this respect extensively used for designing reservoirs and hydrological systems, particularly where the analytical probability methods have so far not been fully elaborated, or where they are completely lacking.

The model is applied most readily to the average annual flow series, the properties of which can simply be interpreted by probability distribution (the first three moment characteristics invariably prove sufficient) and by the autocorrelation function. That is also why the achievement of good agreement between the inputs and the outputs does usually not pose any problems.

The application of the model to the average monthly flows, with which it is essential to follow the probability distribution of the flows in the individual calendar months, as well as correlation between these flows, proves to be far more complex. The achievement of the agreement between the inputs and the outputs is therefore relatively highly conditional upon the estimate of the theoretical distribution fitting as closely as possible the given empirical distribution of the flows in the individual calendar months, or upon adequate transformation being found for the conversion of the given distribution to normal distribution, which is facilitated by the derivation of twelve regression equations. The currently used log-normal transformation of the real flows often raises the problem of the adequacy and fit of the transformation in cases of lower skewness. This is closely linked with the often difficult estimation of the minimum flows of the months; these estimates should correspond to the length of the random sequence assumed.

This problem can be solved with some approximation with the help of graphical extrapolation or a computer using variant analysis based upon the condition of zero skewness of the transformed flows. Routine modelling will also have to adopt genetic points of view of the minimum flows in order that the estimates may approximate as closely as possible to the real conditions of the given river-basin.

From the hydrological point of view, the disadvantage in using a linear regression stochastic model of monthly flows consists mainly in the fact that no success has so far been achieved in introducing into it the parameters of annual flows. The agreement between the inputs and outputs is therefore sometimes hard to achieve. This drawback can prove troublesome particularly when storage reservoirs are to be designed whose long-term components of the volume stored are functions of the statistical parameters of the average annual flows.

An example of the parameters of average annual flows computed from a modelled 1 000-year series of average monthly flows of the river Berounka at Křivoklát is presented in Table 27. Applying the original form of the model, good agreement was on the whole achieved with the first three parameters of distribution, the autocorrelation function does however not correspond to the empirical harmonic function.

Better agreement of the parameters can be achieved if the series of average annual flows and the series of average monthly flows are modelled separately and the latter is then used only as a source of random fragments, which are then assigned to the modelled annual flows according to a suitable rule*).

*) The assignment itself is quite a complex problem in view of the fact that between the annual runoffs and their distribution to the individual months there exist only stochastic relationships. The simplest is the one assigning the average flow of each year a fragment with the same, or nearly the same, annual flow.

TABLE 27. Comparison of the characteristics of a series of average annual flows in the Křivoklát profile of the river Berounka with the parameters of the random series

Series	Characteristics (parameters)									
	$\bar{Q}_a$ ($m^3\ s^{-1}$)	C_v	C_s	$r(1)$	$r(2)$	$r(3)$	$r(4)$	$r(5)$	$r(6)$	$r(7)$
Real series (in the period of 1931 to 1980)	32.646	0.445	1.085	0.543	0.035	−0.115	−0.076	−0.166	−0.417	−0.337
Random 1000-year series (original model)	33.405	0.419	0.973	0.099	−0.035	0.005	0.017	0.005	0.012	0.019
Random 1000-year series (modification)	33.192	0.443	0.938	0.519	0.017	−0.108	−0.044	−0.091	−0.329	−0.243

We explained the principle of this modified model within the framework of our research carried out as part of the national plan of basic research in 1975 [79]. Table 27 shows that a substantial advantage of that model consists particularly in the better fit of the curve of autocorrelation function of the average annual flows.

The procedure proposed has however some drawbacks. Separate modelling of the two series makes excessive demands on computer technology. The application of the principle of fragments as well as the fact that the individual years have different linear transformations cause some impairment of the probability properties of the flow series in the different months; with longer random series linear transformations have however no substantial effect on the properties of the monthly flows, and the variations of the input parameters range mostly within the limits of the admissible random monthly errors.

The recent world-wide tendency is to construct a fitting stochastic model of the flow series with steps not exceeding one month. Intensive attention has particularly been given to the possibilities of deriving a model of the average daily flows, which would be of great help in the solution of important water management problems (viz. short-term equalization of runoffs with the help of reservoirs, compensational runoff control, the effect of man's activities on hydrological regimes etc).

The modelling of average daily flows is a complex task, because these flow series manifest relationships between precipitation and runoffs, and also because an extraordinarily large number of characteristics must be taken account of. These models are therefore used to examine various methodological approaches simulating hydrological regimes under changeable conditions. As one of the oldest can be regarded the classical principle of linear regression usually applied to sequence stripped of the trend and the periodic component. Special models of daily flows are even being developed (e. g. based upon the application of Poisson's processes). The fragments method is often applied with the fragments defined similarly as in the monthly flow series, viz. as real hydrograms of daily flows divided by the average flow of the respective year. The method has several advantages (e. g. retention of the correlations within a year), but also some drawbacks, which have already been discussed.

In the construction of mathematical models of the average daily flows due account must also be taken of considerable variability of these flows, which depends upon a number of factors. For instance, research into these relationships carried out in the Czechoslovakia [43] showed that the variability of the M-daily flows*) was rather changeable, viz. it varied both with variable M and

*) An M-daily flow, Q_{Md}, is the average daily flow reached or exceeded after M days of the period selected. It is determined with the help of the line of transgression of the daily flows plotted for the same period for which the long-term average annual flow has been calculated.

of course regionally. And the variability of the regime of the M-daily flows depends quite clearly upon both the magnitude of the long-term flows, or the depth of the runoffs, and the hydrological structure of the river-basin.

In this field, research has also revealed other interesting facts, viz. that the variability of the M-daily flows in the individual years is different for flows in the region of maxima, for flows of medium volumes, and for the minimum flows in most river-basins.

The development of the investigation of the models of average daily flows in the Czechoslovakia has recently been the subject of Szolgay's paper [109], which also tests in detail the applicability of the method of fragments. The author shows that mathematical modelling of time series is still receiving great attention in many countries. Owing to its considerable importance, this matter seems to be well worth special monographic treatment.

11.3 Modelling of random hydrological series with respect to the bias of the characteristics of the given real sample

The representativeness of the given real series is a serious problem of the application of a linear regression stochastic model. In practice it is currently assumed that the characteristics of a real series have the weight of the parameters of the population, and the model is therefore very often derived directly from the original real flow series; its representativeness thus corresponds to the representativeness of the series.

The results of research however indicate that the assumption of representativeness (in the probabilistic sense of the word) may not generally be satisfied by the real flow series. Their characteristics are very often biassed and they need to be corrected as far as the systematic errors are concerned so that more dependable estimates of long-term parameters may be obtained to be fed into the model's input.

The correction of the characteristics involves a new and difficult task for the methodology of modelling monthly flows, viz. the task of finding algorithms built only on the corrected characteristics, for which the real sequence is unknown, not on any real sequence of flows.

The problem of bias of the characteristics involved in the modelling of random series has been dealt with by a number of authors. The properties of the estimates of autoregression parameters derived with the help of the current method of least squares were examined e. g. by Anděl [2]. Anděl arrived at an important finding, viz. that the method may provide a consistent estimate of the vector of autoregressive parameters, nevertheless the estimates may not generally be unbiassed. At the same time he pointed out that in this respect some authors were sceptical, particularly as far as the application of the method of least squares to short sequences is concerned (viz. those for which $n < 40$); but the order of the sequence is obviously what also matters here.

The same finding concerning the consistency of the estimates of the autoregressive parameters and their possible bias with shorter series was arrived at by Kos [55].

The problem of bias is also subject to Kos's further study [56]. For modelling random series with triparametric log-normal distribution he recommends two numerical procedures. One of these makes use of the exact relationships between the characteristics of the given variables and the characteristics of the transformed variables $y_i = \ln(x_i - x_0)$ (comp. equations (4.3), (4.4) and (4.5), or (4.10), (4.11) and (4.12)). The recurrent relationship for generating the values of y_i is then simply expressed with the help of the respective parameters of variables y_i. Variables y_i are thus generated as normally distributed and they are then converted to a sequence of variables x_i by inverse transformation, viz.

$$x_i = x_0 + \exp(y_i) . \tag{11.20}$$

The other procedure again assumes the triparametric log-normal distribution of variables, for the transformed variables y_i their moment characteristics are however determined directly. The estimates obtained are biassed, even though the deviations may in no way be very large.

This procedure has its advantages. One of these is the possibility of using it for constructing Markovian models of higher orders. And that is also why this procedure was applied in the preparatory studies for drawing up the Directive Hydrological Plan of the Czechoslovakia.

The traditional method of minimum residual variance usually leads to high estimates of the order of autoregression. Although the models derived in this way may be quite a fitting description of the correlation pattern of the given time series, they can be relatively very complex and costly, which is obviously a drawback as far as the generation of synthetic series with the help of a computer is concerned. Other methods have recently come into use, viz. those penalizing the selection of a too high order of the model and simultaneously providing point estimate $\hat{p}$ of the order of the autoregressive model.

As particularly successful can be viewed the methodological procedures which were oriented to the function

$$A_{k,l} = \hat{\sigma}^2_{k,l}[1 + w(k, l)], \qquad k = 0, 1, \dots, K, \; l = 0, 1, \dots, L, \tag{11.21}$$

where K, L denote the predetermined upper limits of the p, q order of the ARMA(p, q) model, and $w(k, l)$ the penalizing function with arguments k, l minimizing the expression (11.21), rather than to the value $\hat{\sigma}^2_{k,l}$ of variance σ^2 of the white noise in the ARMA(k, l) model estimated. The order of the model is thus determined by a compromise between the excessive values of k and l with a low value of variance $\hat{\sigma}_{k,l}^2$ and the low values of k and l with an excessively high estimate of $\hat{\sigma}_{k,l}^2$. With the given length n of the series, the penalizing function must thus be the increasing function of arguments k, l, and with the values of k, l fixed and n increasing, it must converge towards zero.

The literature has several expressions for the penalizing function $w(k, l)$. Substituting these expressions into equation (11.21) and taking logarithms will yield the criteria of estimation, the minimization of which will give us the order of the model sought.

The AIC criterion (Akaik's Information Criterion) has the following form:

$$\mathrm{AIC}(k, l) = \ln \hat{\sigma}_{k,l}^2 + \frac{2(k + l)}{n}, \tag{11.22}$$

where ln stands for the natural logarithm.

The criterion is quite simple and therefore frequently applied. It can however sometimes lead to and overestimation of the order of the model [25].

The FPE (Final Prediction Error) criterion in the following form,

$$\mathrm{FPE}(k) = \ln \hat{\sigma}_k^2 + \frac{2k}{n}, \tag{11.23}$$

is a special case of the AIC criterion. The order of an autoregressive model is determined on the basis of the FPE criterion so that it may give the minimum prediction by one step forward.

The BIC criterion (Bayesian Information Criterion) has the following form:

$$\mathrm{BIC}(k, l) = \ln \hat{\sigma}_{k,l}^2 + (k + l)\frac{\ln n}{n}. \tag{11.24}$$

It was derived independently by Schwarz and by Rissanen [127, 128]. Its estimates are highly consistent.

The HQ (Hannan–Quinn) criterion has the following form:

$$\mathrm{HQ}\,(k, l) = \ln \hat{\sigma}_{k,l}^2 + c(k + l)\frac{\ln\,(\ln n)}{n}. \tag{11.25}$$

It was derived by Hannan and Quinn originally for the autoregressive models, and they proved its high consistency for $c > 1$. Its generalization for the ARMA models was suggested by Hannan, who also proved the consistency of the (11.25) estimate for $c > 2$.

For the determination of the order of the ARIMA(p, d, q) model, Ozaki [89] and also Cipra [25] suggested the criterion AIC(k, d, l) in the following form:

$$\mathrm{AIC}(k, d, l) = \ln \hat{\sigma}^2_{k,d,l} + \frac{2(k + l + 1 + \delta_{d0})}{n - d}, \tag{11.26}$$

where δ_{d0} is Kronecker's delta, i. e.

$$\delta_{d0} = 1 \text{ for } d = 0\,,$$
$$\delta_{d0} = 0 \text{ for } d \neq 0\,.$$

In the Czechoslovak water-engineering literature these problems were dealt with by Procházka [91], who clarified the principle of penalization and analyzed several methods of estimating the autoregression order, which he compared with the method of minimum residual variance in thirty series of average monthly flows. He then drew the conclusion that in view of the dependability of hydrological data that can be achieved, the upgrade of residual variance with the help of the method of least squares was practically negligible and the corresponding increase of the order of autoregression useless. He therefore recommended using the more modern methods of estimating the order of autoregression making use of the penalizing function and leading to lower and more efficient estimates.

In our research we tested three numerical methods of generating random series of average monthly flows with respect to bias. In the first test, we first modelled a series of average annual flows with estimated unbiassed parameters*), which we then assigned the random fragments of monthly flows obtained with the help of the classical linear regression model. The experiment was only partly successful, for satisfactory agreement was achieved only with the long-term average monthly flows and their coefficients of variation. The greatest difficulties were posed by the skewness of the monthly flows, which is oversensitive to the input values of parameters.

The second series of experiments was based upon the utilization of the correlation functions between vectors of the average monthly flows in the pair of the neighbouring months. For the average monthly flows their unbiassed parameters were estimated first. As the next step, the flows were modelled under the assumption of their independence. The mutual correlation functions were then computed, and such outliers among the vectors of the monthly flows were sought in their curves that manifested relatively best agreement with the empirical relations of correlation. Neither of these experiments was fully successful, because the relations of correlation between the flows of the neighbouring months were mostly statistically insignificant, which could have been expected.

---

*) For modelling annual flows use can be made of the algorithms quoted e. g. in Soviet literature [96, 107, 108].

In the third series, we started with the theoretical relationships between the characteristics of the given series of monthly flows quoted above and the characteristics of their logarithms. The characteristics of the real series of the average annual and monthly flows had to be corrected as far as the systematic errors are concerned, which provided us with more reliable input parameters for the model.

If the routine solving procedure is applied, the characteristics of the real series are corrected as follows:

- the given expected real values (assumed to be unbiassed) can directly substitute for the estimated long-term expected values of flows in the individual calendar months;
- the sample coefficients of variation and asymmetry of the monthly flows are corrected as far as their systematic errors are concerned (by the systematic errors being added to the values of these coefficients);
- the coefficients of correlation of the real series are directly substituted for the estimates of the long-term unbiassed values of the coefficients of correlation between the flows in the neighbouring months; in agreement with the results of the research, the systematic errors are assumed to be insignificant, even negligible;
- from the parameters of the series of the average annual flows the coefficients of variation and asymmetry are subjected to correction (with the long-term expected values assumed to be unbiassed).

As the next step, the series of annual and monthly flows are modelled for the inputs estimated, as already mentioned above in Section 11.2. In the individual months the monthly flows modelled are then regarded as random fragments, which are then assigned to the modelled series of annual flows.

As the last step, the measure of agreement between the input and the output parameters is checked, invariably at the level of significance of 5 percent.

The reader will find an example of the results achieved by the modelling of a random series of average monthly flows with respect to the bias of the characteristics of the given real sample in Table 28, which compares the estimated inputs and outputs of the random series in the Děčín profile of the river Elbe. Statistically significant deviations were found in only two cases out of the fifty-two parameters of the series checked, which can be considered a satisfactory result.

This modification of the linear regression model has upgraded the original qualities of the model in two ways: firstly, using the model can help to establish satisfactory agreement between the annual parameters, which are of decisive importance particularly as far as the design of storage reservoirs is concerned; and secondly, apart from this, modelling can take account of the bias of the characteristics of the real sample and help to obtain more reliable estimates of the input parameters.

TABLE 28. Survey of the estimated and output parameters of a 1000-year random series in the Děčín profile of the river Elbe

	Parameter estimated (inputs)				Outputs of the random series			
	$\bar{Q}$ (m^3 s^{-1})	C_v	C_s	$r(1)$	$\bar{Q}$ (m^3 s^{-1})	C_v	C_s	$r(1)$
years	311.8	0.31	1.15	0.344	315.0	0.312	1.137	0.325
XI	233.8	0.64	2.36	0.600	239.2	0.686	2.363	0.038*)
XII	281.9	0.70	2.61	0.625	290.2	0.721	2.343	0.679
I	324.8	0.75	2.92	0.444	332.8	0.845	3.731	0.496
II	387.4	0.59	1.29	0.374	392.1	0.809	1.626	0.407
III	529.8	0.52	1.83	0.205	536.0	0.529	1.769	0.203
IV	506.4	0.55	2.14	0.553	517.5	0.597	2.284	0.466
V	357.1	0.51	2.28	0.488	360.5	0.582	3.128	0.518
VI	265.0	0.76	4.77	0.422	258.9	0.721	4.030	0.447
VII	238.1	0.70	2.68	0.442	235.5	0.705	2.573	0.362
VIII	201.7	0.69	3.12	0.418	200.2	0.740	3.517	0.524
IX	196.5	0.70	2.36	0.615	195.1	0.752	2.305	0.709
X	219.3	0.62	2.89	0.449	222.1	0.683	2.992	0.668*)

*) Statistically significant deviations at the 5 % level.

The numerical method described above has however some drawbacks. As compared with the classical form of the regression model, the number of computing operations has considerably increased, which of course prolongs the machine time needed to deal with the overall task. The relatively time-consuming ascertainment of the systematic errors with the help of an analysis of the set of random samples can be substantially facilitated by computer-aided automatic estimation of the parameters (see Section 11.1).

Modelling random series with respect to the bias of the sample characteristics is another step towards more adequate processing of hydrological data. But what still needs to be studied is the methodology of modelling the average monthly flows in the system of stations with respect to bias.

12
Application of the theory of estimation to the design of storage reservoirs

12.1 Long-term stationary function of storage reservoirs

The importance of the representativeness of hydrological data to the design of reservoirs has been proven by a number of studies and scientific treatises. In spite of the results achieved, no unified and formalized procedure of estimating unbiassed parameters of flow series, upon which the design of reservoirs is based, has so far been devised. This leads to repercussions in the probability methods of computing the storage function of reservoirs. In the current practice of water-engineering computations this drawback manifests itself in the fact that satisfactory representativeness of the initial flow series is invariably assumed without the dependability of the result of the hydrological solution being subjected to any examination.

However, the lack of any methodological procedures oriented towards the correction of the biassed characteristics of the real flow series also manifests itself in the modelling of random series. The directly computed characteristics are often treated as inputs of models, and with the parameters of the modelled series (outputs) their agreement with the inputs is to be expected, except for the statistically insignificant deviations.

Under these circumstances the advantage of the probability methods of designing reservoirs over the solution with the help of short real flow series consisted in the fact that the function of reservoirs was more reliably expressed with the help of sufficiently long (theoretically infinitely long) stochastic sequence of inflows into the reservoir. But the representativeness of this sequence in the sense of probability corresponded only to the original real series.

The long-term stationary function of reservoirs can well be dealt with with the help of the random flow series modelled with respect to the input parameters desired. The advantage of this method as compared with the analytical approaches consists primarily in the fact that the function of a reservoir can practically be formulated for any arbitrarily selected properties of the flow series, including their autocorrelation functions.

In our research [87] we computed the storage function of reservoirs with the help of average monthly flows, the long-term representative parameters of which we estimated by adhering strictly to the principles of the estimation theory. For the real flow series selected, we first analyzed the measure of bias of the statistical characteristics related to the length of the sample and we derived the systematic errors, which we used for correcting the sample characteristics. The parameters estimated in this way were then brought on to the inputs of the models of the 1 000-year series of average monthly flows, which were thus employed in the computation of the storage function of reservoirs.

In harmony with the results of our previous research, compensation was undertaken of the coefficients of variation and asymmetry of the real flow series, to which systematic errors were added. The estimates of the long-term expected values of the flows in the individual calendar months were substituted by the given expected real values (the unbiassedness of which can be proved). And the estimates of the long-term unbiassed values of the coefficients of correlation between the flows of neighbouring monthly flows were also substituted by the coefficients of correlation of the real series themselves (the systematic errors of which are next to negligible).

The effect of the bias of the statistical characteristics of the flow series upon the computation of the storage function was assessed with the help of a comparison of the approach to the design of reservoirs based upon random series with parameters equal to the given characteristics of the real series, with the approach based upon random series with the characteristics corrected to unbiassed parameters, i. e. with systematic errors added. In the first case our point of departure was the assumption hitherto currently adopted in the modelling of random series, viz. that in the model, the characteristics of the original real series have the weight of parameters. We then compared the two approaches with the approach based on the original real series.

In order that the effect of the bias of the characteristics might be assessed with the seasonal and long-term controlled runoffs, we opted for a relatively wide interval of the coefficients of minimum plus runoff α(α = 0.3; 0.4; 0.5; 0.6; 0.7; 0.8; 0.9) and specific sizes of the storage volumes of reservoirs β (β = 0.10; 0.20; 0.30; 0.40; 0.50; 0.60; 0.75; 1.00; 1.25; 1.50; 1.75 and 2.00). For the pairs of the values of α and β (in the real series and in the two variants of the random series) we sought the insurances of the discharge of water with respect to repetition p_o, duration p_t, and supply p_d.

The results of the numerical solution of a large number of variants were graphically represented in the form of the currently used curves of relationships $\beta = f(\alpha, p_o)$, $\beta = f(\alpha, p_t)$, and $\beta = f(\alpha, p_d)$. A comparison of these curves made it possible to characterize the effect of the bias of the characteristics of the flow series upon the computation of the storage function of reservoirs.

We compared the results arrived at with the help of the real series with those

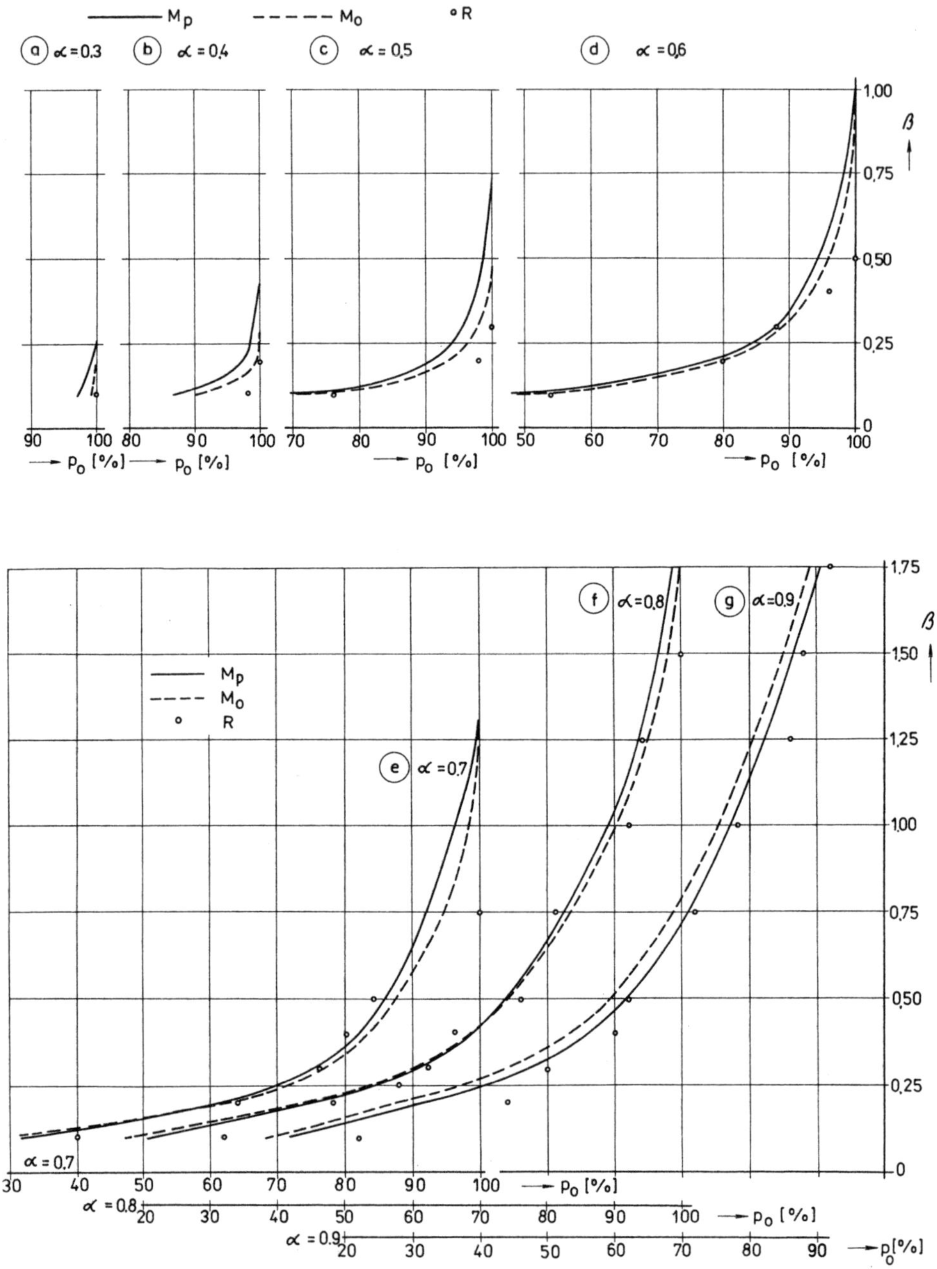

Fig. 48. Regime curves of the total storage volumes of the reservoir, $\beta = f(\alpha, p_0)$, in the Křivoklát profile on the river Berounka: M_p – from the solution using the original 1 000-year series, M_0 – from the solution based upon the corrected 1 000-year series, R – the values from the solution on the basis of the real series, p_0 – insurance with respect to repetition (%).

obtained in the case of the two random series. From Fig. 48 it can be seen that the former invariably (except for $\alpha = 0.7$ and $\alpha = 0.8$) lead to lower demands concerning the storage volume than the latter. Using the real series can thus lead to unreliable results, which the theory of reservoir-runoff control accounts for by the well-known fact that the given flow series can well be of an absolutely random character.

The relationships between the storage volumes and the insurance of runoff compensation are also in keeping with the contemporary knowledge derived nearly twenty years ago by a comparative analysis of the results of the computation of the storage capacity of reservoirs with the help of real series and the solution of that problem with the help of diagrams of the long-term components of reservoirs [116]. It turned out that even with long-term runoff control ($\alpha = 0.8$–0.9) the results of the solution achieved with the help of the real series and those obtained using random series might not differ so markedly in a number of cases.

The relationship between the M_p and M_0 curves presented in Fig. 48a) to g) is characteristic: with seasonal and long-term runoff control curve M_0 invariably lies below curve M_p, so that determining the storage with the help of the corrected modelled series will be more economical, the only exception being the markedly long-term runoff control ($\alpha = 0.9$ in Fig. 48g), where higher required storage capacities are obtained for equally ensured runoffs with respect to repetition after the statistical characteristics of the average annual and monthly flows have been corrected. The same tendency can be witnessed if a comparison is undertaken of the curves of ensured runoff with respect to duration. As far as runoff insurance with respect to the volume supplied is concerned, the absolute and relative differences between the ordinates of the two curves become markedly less pronounced, and with α achieving higher values that tendency will fade completely.

In commenting on the mutual position of the M_p and M_0 curves of relationship we must base our explanation upon the absolute and the relative measure of the correction of the statistical characteristics of the average annual and monthly flow series in the individual months. It can generally be claimed that a rise in the value of C_v makes the requirements concerning the storage volume of the reservoir rise too. This relationship is particularly evident as far as the long-term component of the storage capacity is concerned; it can however also be related to the total storage capacity, with the seasonal component considered, in spite of the fact that the effect of C_v and its systematic error in the individual months is undoubtedly peculiar. As it turns out, the difficulty of the assessment of the effect of the individual characteristics on the required size of the storage volume will not be mitigated even though C_v of the chronological series of all the average monthly flows may be used.

Raising the value of C_s by adding the systematic error can on the other hand lower the requirements for the storage capacity of the reservoir. The sensitivity of this capacity to the skewness of the probability distribution may be lower than the sensitivity of that volume to the variability of that distribution, but in view of the invariably absolutely higher values of C_s for both the years and the months, its effect upon the results can be quite considerable. This is exemplified by the mutual relationship between the M_p and the M_0 curves for the river Berounka at Křivoklát.

For the values of α up to 0.8 the solution making use of the corrected random flow series leads to lower and more economical results. The measure of C_s correction to an unbiassed estimate is substantially more pronounced than C_v's as far as both the absolute (uncomparable) and the relative (mutually comparable) values are concerned. With the annual flow series, $C_v = 0.44$ was corrected by $\Delta C_v = 0.02$, i. e. approximately by 4.5 %, whereas the correction for $C_s = 1.08$ amounted to $\Delta C_s = 0.37$, i. e. approximately 34.2 %. The correction of C_v of the average monthly flows in the individual calendar months (the original values ranging between 0.58 for Decembers and 0.93 for Julys) of 1.3 only (for Augusts) to 8.1 percent (for Novembers) was again relatively low as compared with the analogous correction of C_s. The initial values of C_s ranged between 0.99 for Februarys and 3.08 for Junes, and their corrections amounted to 21.7 percent for Junes and 62.7 percent for Novembers. These systematic errors of C_s of the average monthly flows have a positive effect on the lowering of the storage volume of the reservoir.

The values of the storage volume required for effective long-term runoff control according to the method of the solution making use of the corrected series (M_0) are higher than the values of these volumes according to the method using original modelled series (M_p). The size of the former is accounted for by the more pronounced effect of the correction of C_v of the series of average monthly flows as compared with the more moderate change of C_s of the same series in the region where the effect of the long-term component of the storage volume of the reservoir prevails, and the seasonal component is of minor influence only.

Interesting results were also provided by the water-engineering solutions of the problem of the storage capacity of reservoirs when use was made of other characteristic flow series relative to the Elbe river-basin.

The river Mrlina at Vestec is characterized by extraordinary fluctuation (the annual flow series exhibiting C_v equal to 0.645). This corresponds to marked differences between the solutions of the problem of the storage volume of reservoirs in different series (R, M_p, M_0). The respective curves $\beta = f(\alpha, p_o)$ of runoff insurance according to repetition, analogous to those relating to the river Berounka at Křivoklát, are presented in Fig. 49a) to g).

The initial real flow series of the Mrlina at Vestec was the shortest (viz. only 26 years) of the set of the series examined, which was mirrored in the rather different shapes of regime curves R and M_p or M_0, particularly in the region of

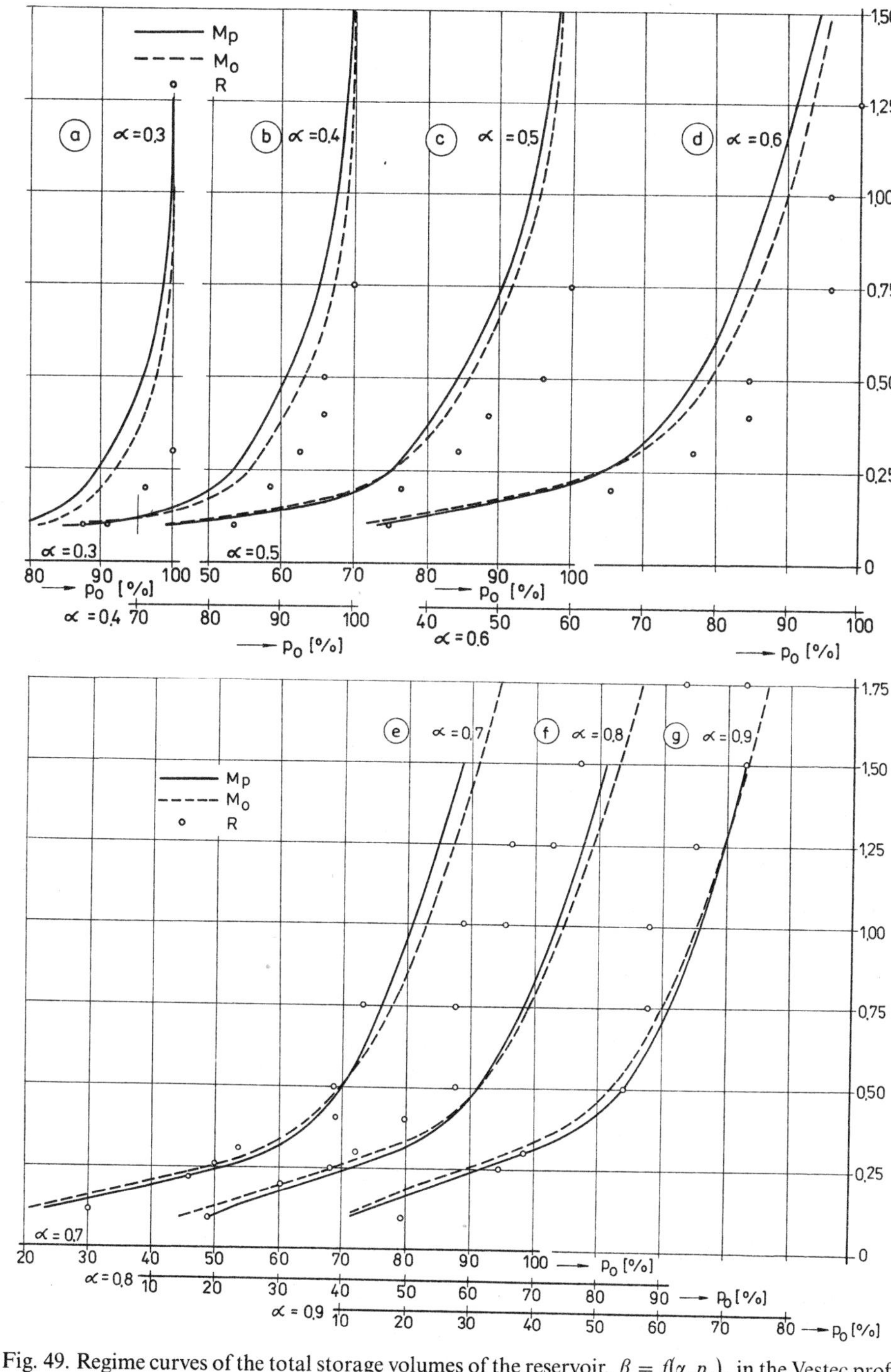

Fig. 49. Regime curves of the total storage volumes of the reservoir, $\beta = f(\alpha, p_o)$, in the Vestec profile on the river Mrlina: M_p – from the solution using the original 1 000-year series, M_0 – from the solution based upon the corrected 1 000-year series, R – the values from the solution on the basis of the real series, p_o – insurance with respect to repetition (%).

the transition from seasonal to long-term control ($\alpha = 0.5$ and 0.6; Fig. 49c), d)), but also in the region of seasonal control alone ($\alpha = 0.3$ and 0.4; Fig. 49a), b)). In the case of long-term control the differences in the shapes of the curves become less marked, and for $\alpha = 0.8$ and 0.9 some sections of curve R are even above the M_0 and M_p curves. This can be accounted for by the fact that the short real series comprised an extremely dry period, the probability of the occurrence of which should have been related to a substantially longer historical period. The shapes of the regime curves for $\alpha = 0.8$ (Fig. 49 f), g)) do however not do justice to the region of the highly ensured runoffs of roughly more than 85 % for $\alpha = 0.8$ and exceeding 80 % for $\alpha = 0.9$, where the relationship between the curves can acquire a different shape.

In view of a higher variability of the flows of the river Mrlina at Vestec, the absolute differences of the ordinates of curves M_p and M_0 are usually more pronounced than for instance the flows of the river Berounka at Křivoklát. (In Figures 48 and 49 the curves are plotted on the same scale). The differences are again absolutely and relatively the highest in the region of the transition between the seasonal and the long-term runoffs.

The most important finding is that the solution of the problem of the storage

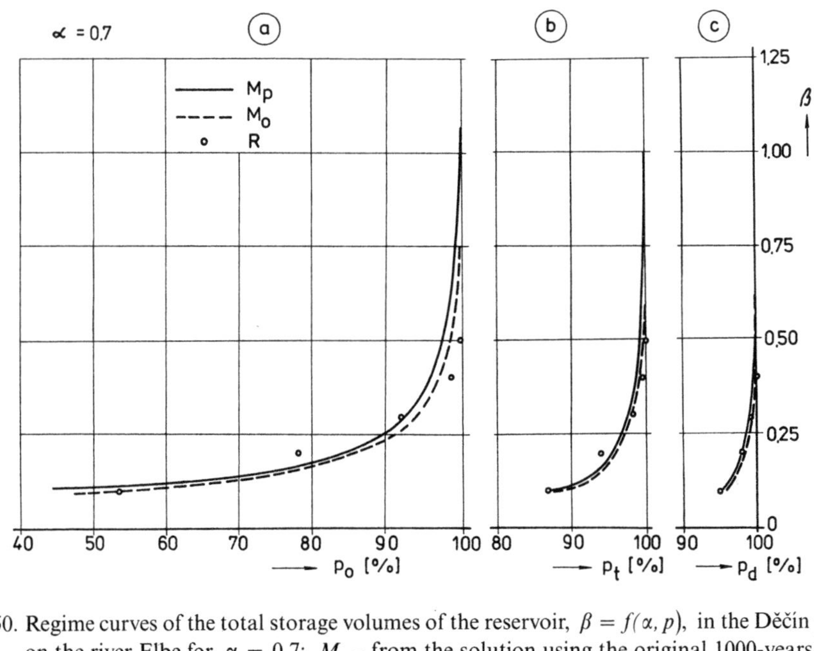

Fig. 50. Regime curves of the total storage volumes of the reservoir, $\beta = f(\alpha, p)$, in the Děčín profile on the river Elbe for $\alpha = 0.7$: M_p – from the solution using the original 1000-years series, M_0 – from the solution based upon the corrected 1 000-year series, R – the values from the solution on the basis of the real series; (a) for insurance with respect to repetition p_o, (b) for insurance with respect to duration p_t, (c) for insurance with respect to the volume of the water delivered p_d.

volume with the help of the corrected flow series can in a number of cases be more acceptable, viz. economical.

The water-engineering solution of the problem of the storage function of reservoirs with respect to the bias of the statistical characteristics of the flow series can in some cases lead to further interesting results, which can be expressed with the help of curves $\beta = f(\alpha, p)$.

Thus, for instance, in Fig. 50 the differences between curves M_p and M_0 for the Elbe at Děčín with a ninety-year original real series are invariably smaller than in the preceding cases. This is accounted for by the fact that the lower values of C_v and C_s also correspond to lesser systematic errors, so that the probability properties of the two random series are closer to each other. The example of the shape of the curves for $\alpha = 0.7$ selected shows that smaller differences in the ordinates of curves $f(\alpha, p_o)$ also correspond to smaller differences in the ordinates of curves $f(\alpha, p_t)$ and $f(\alpha, p_d)$.

The flow series of the river Jizera at Vilémov is characterized by very low variability (the annual flows exhibiting C_v equal to 0.234) and near-zero asymmetry (C_s = 0.017).This also corresponds to some specific properties of curves $\beta = f(\alpha, p)$.

Figure 51 shows the characteristic shape of these curves for $\alpha = 0.8$, again for all the types of runoff insurance, p_o, p_t, p_d. As in other cases, the solution making

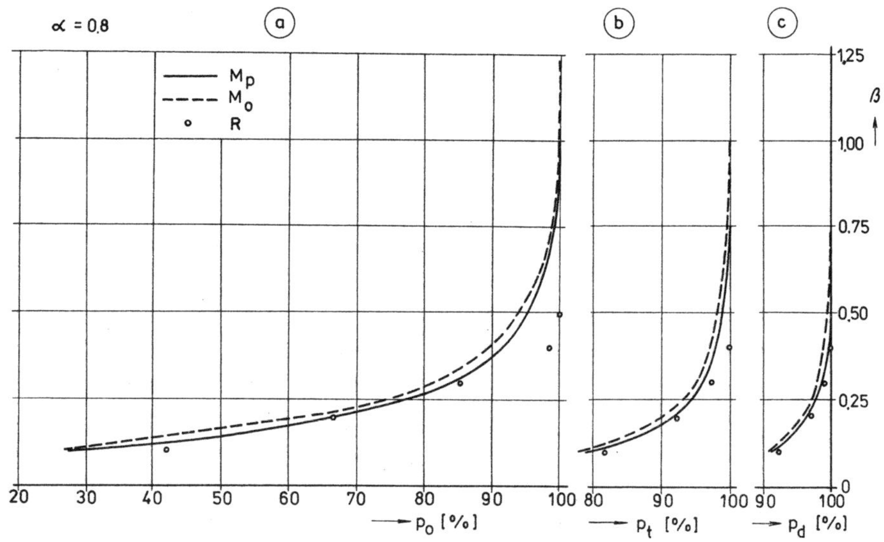

Fig. 51. Regime curves of the total storage volumes of the reservoir, $\beta = f(\alpha, p)$, in the Vilémov profile on the river Jizera for $\alpha = 0.8$: M_p – from the solution using the original 1 000-year series, M_0 – from the solution based upon the corrected 1 000-year series, R – the values from the solution on the basis of the real series; (a) for insurance with respect to repetition p_o, (b) for insurance with respect to duration p_t, (c) for insurance with respect to the volume of the water delivered p_d.

use of the real series leads to less dependable storage volumes, particularly as far as the higher runoff insurance with respect to repetition, duration and the volume of the water supplied is concerned. Great interest is of course aroused by the mutual relationship of curves M_p and M_0, the shapes of which approach each other even with relatively high values of runoff. The mechanism of the effect of the systematic errors of the individual characteristics of the annual and monthly flows upon the solution of the problem of the storage capacity of reservoirs is obviously rather complex in this case, and the positive effect of the systematic errors of one parameter (e. g. C_s), which tend to reduce the size of the storage volume, can off-set the adverse effect of the systematic errors of another parameter (e. g. C_v). The mutual compensation of the effects of various parameters on the solution of the problem of the storage function of reservoirs reminds us of the effect of the positive and negative ordinates of the autocorrelation function of the average annual flow series on the magnitudes of the long-term components required [83].

Assessing the effect of the individual characteristics of the average monthly flows, as well as their systematic errors, on the required magnitude of the storage volume, is thus an extraordinarily difficult task even now, when the theory of mathematical modelling of hydrological series has reached a considerably high level of elaboration. If we consider, for instance, the fact that the linear regression stochastic model of the average monthly flow series is entered by more than fifty characteristics, then it stands to reason that the effect of these characteristics on the solution of the problem of the storage volume can practically be approached only summarily via a random series, the parameters of which differ from the input characteristics by only statistically insignificant deviations.

A closely related and rather complex problem is also the problem of the mutual relation of the random deviations in a pair of random series modelled with the estimated unbiassed parameters and with parameters equal to the characteristics of the given real series. As is well-known, the output parameters of even the long modelled series are burdened with certain random deviations. And each combination of the input parameters can correspond to slightly different random deviations in the output. Their differences, derived from the two modelled series, thus affect the precision of the formulation of the effect of the systematic errors on the solution of the problem of the storage capacity of a given reservoir.

The mutual relationships of the pairs of curves M_p and M_0 type $\beta = f(\alpha, p)$ can be supplemented by deviations of the type $\Delta p = f(\alpha, \beta = \text{const})$, which testify particularly to their dependence on the level of runoff equalization with the help of reservoirs under various hydrological conditions. In Fig. 52 they are expressed for four selected profiles of the magnitudes of deviations Δp_0 in the order starting with their highest negative values and ending with their average highest positive values.*)

Deviations Δp in the solutions using series M_p and M_0 were computed for discrete values of the coefficient of minimum-plus runoff $\alpha = 0.3$ to 0.9 with a 0.1 step, and for the relative total storage volumes of reservoirs $\beta = 0.10$; 0.20; 0.30; 0.40; 0.50; 0.75; 1.00; 1.25; 1.50; 1.75; 2.00. The points of the curves of deviations found were linked with a broken line. This approximation does not bias the conclusions.

In the flow series of the Jizera at Vilémov with low variability and near-zero asymmetry of average annual flows (and also low variability and asymmetry of

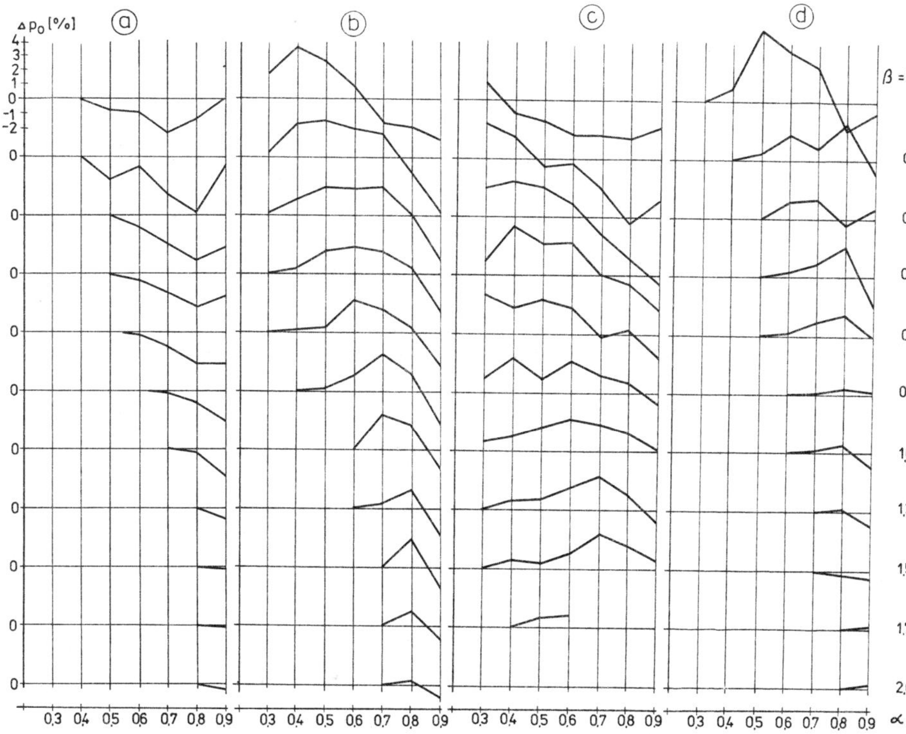

Fig. 52. Differences between the values of the sections of regime curves from the solutions based upon the corrected and the original 1 000-year series $\Delta p_o = p_{opr} - p_{p\mathring{u}v}$ for combinations α and β solved: Δp_o – differences between the values of insurance with respect to repetition; (a) the river Jizera at Vilémov, (b) the river Berounka at Křivoklát, (c) the river Mrlina at Vestec, (d) the river Elbe at Děčín.

*) The order of the profiles is only informative, because a more exact comparison of the curves of deviations p_o, the magnitude of which depends upon a number of characteristics and their systematic errors, is very difficult.

average monthly flows) we found exclusively negative deviations $\Delta p_o = p_{opr} - p_{p\mathring{u}v}$, so that with a certain magnitude of the storage volume of reservoir β the minimum-plus runoff (with respect to repetition) is less guaranteed in the corrected series than in the original modelled series. The maxima of these negative deviations lie rather in the region of the markedly long-term controlled runoffs (e. g. with $\beta = 0.10$ for $\alpha = 0.7$; with $\beta = 0.20$ for $\alpha = 0.8$ etc.); with β rising, they shift in the direction of higher values of α, the absolute values of deviations Δp simultaneously falling off until they become next to insignificant approximately with $\beta \geqq 1.25$.

The other profiles monitored exhibited mostly positive deviations Δp_o, particularly with the low and the average values of the coefficient of reservoir-controled runoff α, viz. prevailingly with seasonally-controlled runoff. In these cases the correction of the input characteristics as far as the systematic errors are concerned leads to more economical solutions of the storage function of reservoirs.

The highest positive deviation was found in the case of the Děčín profile of the Elbe ($\Delta p_o = +5$ % with $\beta = 0.10$). With $\beta \geqq 0.75$ the effect of systematic

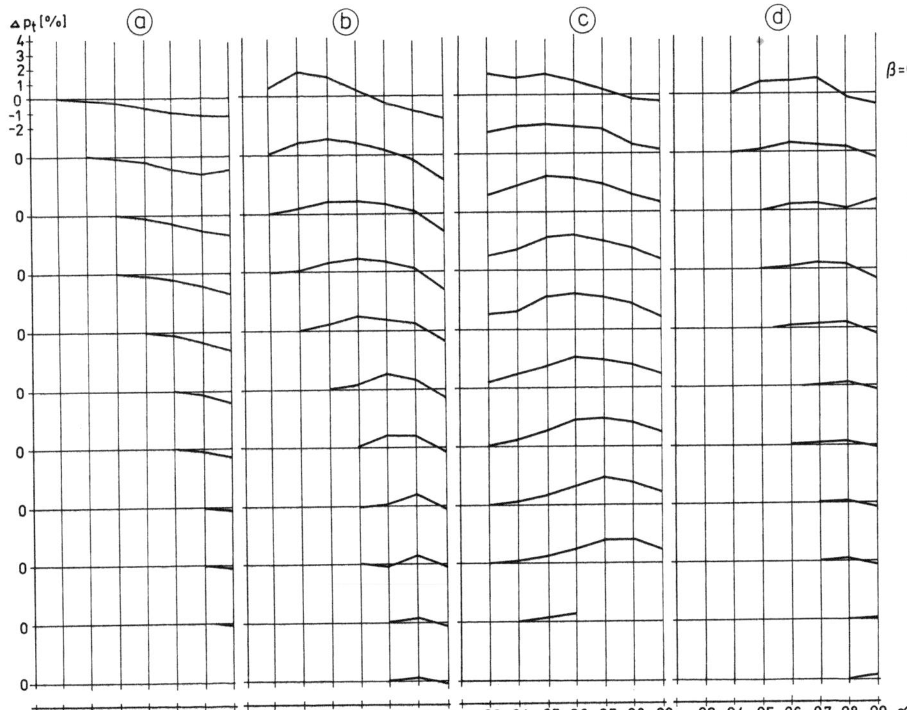

Fig. 53. Differences between the values of the sections of regime curves from the solutions based upon the corrected and the original 1 000-year series $\Delta p_t = p_{opr} - p_{p\mathring{u}v}$ for combinations α and β solved: Δp_t – differences between the values of insurance with respect to duration; (a) the river Jizera at Vilémov, (b) the river Berounka at Křivoklát, (c) the river Mrlina at Vestec, (d) the river Elbe at Děčín.

errors on the design of the storage volume of reservoirs is practically insignificant in that profile. In these cases minor negative deviations Δp_o also occured.

In Fig. 52 another interesting tendency manifests itself in the region of positive deviations Δp_o. In the individual profiles, with β increasing the maxima of deviations Δp_o invariably shift towards higher values of α. The absolute magnitudes of deviations Δp_o fall off with both β and α growing, which is quite logical.

In Fig. 53 we have plotted the dependence of deviations Δp_t upon equally chosen values of β and α for the same four profiles. It can be seen that the curves of this dependence are similar to those of deviations Δp_o. The mostly negative deviations Δp_t are again exhibited by the Jizera at Vilémov; their minima (approximately − 1.4 %) are also in the region of the lower values of β and higher values of α.

The Berounka at Křivoklát shows prevailingly positive deviations Δp_t (max Δp_t = 1.8 %). Their tendency is similar to those of deviations Δp_o.

The highest positive deviations Δp_t occurred on the Mrlina, where they reached values of approximately 2 % even for higher β's. With the exception of

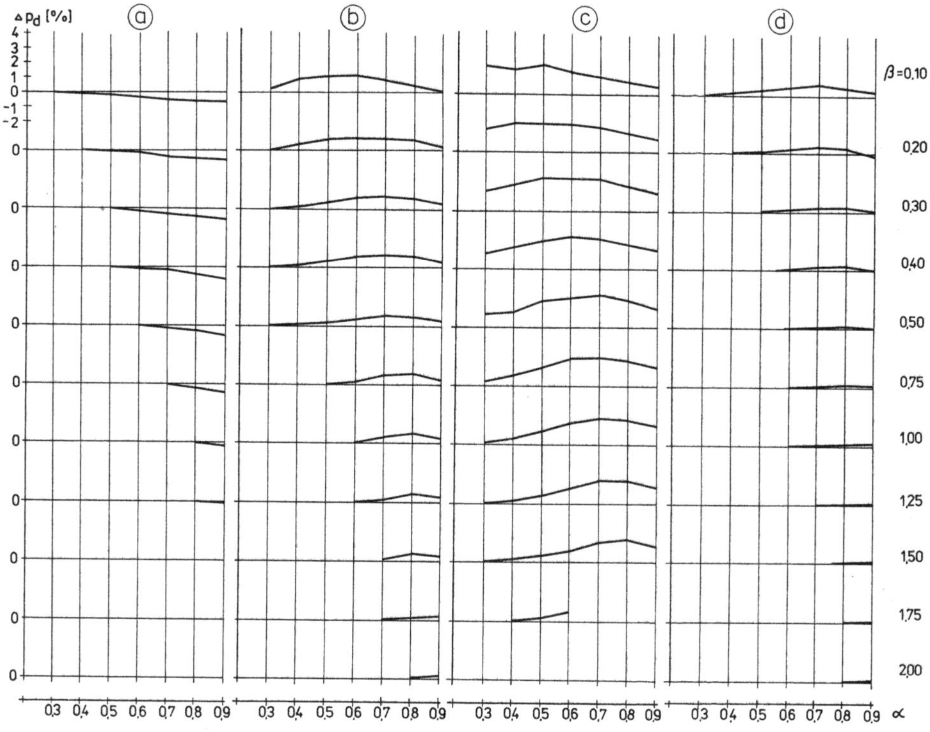

Fig. 54. Differences between the values of the sections of regime curves from the solutions based upon the corrected and the original 1 000-year series $\Delta p_d = p_{opr} - p_{pův}$ for combinations α and β solved: Δp_d – differences between the values of insurance with respect to the volume of the water delivered; (a) the river Jizera at Vilémov, (b) the river Berounka at Křivoklát, (c) the river Mrlina at Vestec, (d) the river Elbe at Děčín.

$\beta = 0.10$, the positive deviations Δp_t occur throughout the whole interval of coefficient α monitored.

On the other hand, the lowest positive values of Δp_t occurred on the Elbe at Děčín, where they quickly decrease to practically negligible values with β growing.

Figure 54 presents a plot of the dependence of deviations Δp_d upon β and α. Its tendency is analogous to the tendency of deviations Δp_o and Δp_t and thus does not require any particular explanation. Maximum deviations: max. $\Delta p_d = + 2.2$ % (the Mrlina), min $\Delta p_d = -0.8$ % (the Jizera).

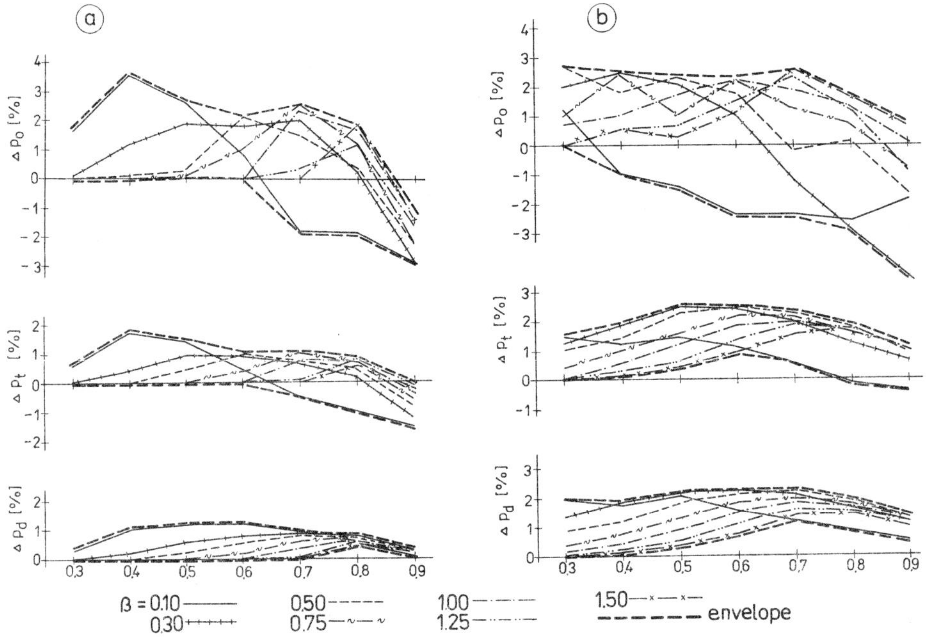

Fig. 55. Differences between the values of the sections of regime curves from the solutions based upon the corrected and the original 1 000-year series $\Delta p = p_{opr} - p_{p\mathring{u}v}$ for various α's with the envelope of the curves of deviations for various β's marked; (a) the river Berounka at Křivoklát, (b) the river Mrlina at Vestec;
Δp_o – deviations of the values of insurance with respect to repetition,
Δp_t – deviations of the values of insurance with respect to duration,
Δp_d – deviations of the values of insurance with respect to the volume of the water delivered.

From the five profiles monitored, Fig. 55 presents Křivoklát on the Berounka and Vestec on the Mrlina, for which we have redrawn deviations Δp_o, Δp_t and Δp_d into joint patterns of all the β's monitored. This type of representation thus gives a summary idea of the tendencies of the values of the deviations dependent upon both β and α. All the patterns also highlight the envelopes of the sets of curves, which give evidence of the fact that the deviations of the guaranteed supplies of water are in no way negligible and that estimation of systematic errors and unbiassed parameters of the flow series is thus of considerable importance for the design of reservoirs.

The hydrological computation of the design of reservoirs in five characteristic flow series of the Elbe river-basin gave the following maximum changes in the guarantee of the reservoir-supplied minimum-plus runoffs, after the respective statistical characteristics had been corrected:

+5.0 %; −4.5 % in the repetition-based guarantee,
+2.0 %; −1.4 % in the duration-based guarantee,
+2.2 %; −0.8 % in the volume supplied-based guarantee.

With the inverse problem the requirements concerning the storage capacity of a reservoir may vary by up to several tens percent.

The knowledge gained from the hydrological computation of the storage function of reservoirs with respect to the bias of statistical characteristics shows relatively considerable sensitivity of the design of the volume of the reservoir to the characteristics of the hydrological regime and their random changes. This sensitivity can manifest itself to various extents, according to the values of the individual characteristics of the flow series, the parameters of the reservoir and the guarantee of the supply of water.

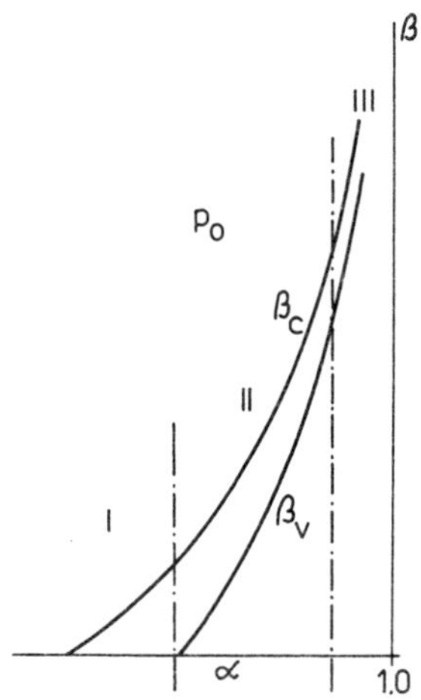

Fig. 56. Schematic representation of a pair of regime curves of total storage volumes of the reservoir and their long-term components.

Under the given hydrological conditions the sensitivity of the computations of the storage capacity of reservoirs to random variations of those conditions can summarily be analyzed with the help of the currently used curves of the type $\beta = f(\alpha, p)$ plotted for the total storage volumes and long-term components.

Figure 56 presents a schematic example of two of these curves, the range of the validity of which is divided into three regions: region I – the total storage capacity is accounted for only by the seasonal component (seasonal runoff control); region II – the total storage capacity is accounted for by both the long-term and the seasonal components (long-term runoff control); region III – the total storage capacity is accounted for predominantly by the long-term component (markedly long-term runoff control; under Czechoslovak conditions approximately for $\alpha > 0.8$).

Region I – the region of seasonal runoff control – is relatively sensitive to the effect of systematic errors of average monthly flows, which can lead to more economical designs of the storage volumes of reservoirs as compared with the solution making use of the flow series with biassed characteristics. With the given seasonal volume of a reservoir the dangerous effect should however not be disregarded of the random variations of hydrological conditions, which can result in a failure of the reservoir to discharge its function (viz. in short operating cycles of the reservoir lower flows, and random variations of these lower flows, can manifest their negative effect). In this region of runoff control it is therefore essential that attention should be given to both the systematic and the random errors of the estimation of the parameters of monthly flows.

In region II a reliable estimate of the parameters of average monthly and annual flows is of the utmost importance. From Fig 56 it can be seen that the mutual relationship between the long-term and the seasonal components varies with p_0 given with respect to α; with α increasing the seasonal component often decreases. The reliability of the computation of the storage capacity therefore depends upon reliable calculation of its two components. The mutual effect of the systematic errors of the parameter estimates of the annual and the monthly flows series is rather complex, and as we already mentioned above, they offset each other.

Region III is characterized above all by the fact that a large increment of β can correspond to a small change of α, particularly under very unsteady hydrological regimes and high design insurance. In this regime, for instance, a reliable estimate of the shape of the correlation function of the annual flows may prove very positive; the non-stationary tendencies of the hydrological regimes, on the other hand, may turn out to be negative [73, 86]. Under these conditions it is thus essential that the economy of the design of the reservoir should be given very close attention.

12.2 Designing storage reservoirs using sets of short realizations of flow series

The assumption of the long-term stationary function of reservoirs has been governing the development of probability methods of hydrological computations since the beginning of this century. The changeable properties of various samples of the same series have however recently raised the question whether the assumption of long-term stationarity of the flow series is so fully justified and uniquely correct as far as the design of reservoirs in limited periods is concerned.

Without the difficult problems of non-stationarity of hydrological regimes and the effect of this non-stationarity on the storage function of reservoirs being considered [73, 86], the answer to that question can safely be sought in the practical aspects of the future operation of reservoirs. What we are interested in most is the near-future period of operation of a reservoir, twenty to thirty years at most, for which we are well capable of estimating the development of the balance-sheet requirements concerning its utilization, taking into account a reasoned forecast of the effect of man's activity, the surroundings of the reservoir, and other conditions upon its function. It follows that it is in this period that we will also be interested in the properties of the hydrological regime, its random variation and effects upon the assumed function of the reservoir.

It also follows that a corresponding hydrological base will have to be selected for computing the storage function of the reservoir in this proximate period. With the prognosis of the future hydrological conditions so unreliable, that base will undoubtedly be the whole set of shorter flow series, which will simulate the possible variations of the hydrological regime in the period under examination.

The idea of using a set of shorter flow series instead of a single sufficiently long random flow series to tackle the storage function of reservoirs in the proximate period appeared as far back as the early 1970s, particularly in the context of the work on the basic material for the Guiding Hydrological Plan (Kos [57]). The matter was also put on the agenda of IInd Symposium on the Methods of Reservoir-Controlled Runoffs held in 1974 at the Faculty of Civil Engineering of the Czech Technical University in Prague (Broža, [18]), where Nacházel [72] presented an analysis of the open problems of the new methodological approach. He concentrated in particular upon the following issues:

1. the relationship between the statistical characteristics of the set of realizations and the statistical parameters of a long random series,
2. the relationship between the runoff insurance in the set of shorter flow series and the design insurance in a long random series,
3. the design values of the storage volume (or minimum-plus runoff for the given volume) computed solely on the basis of a set of realizations.

Contemporary estimation theory is fully capable of coping reliably with the first of these issues. It can be said that this theory, together with sampling theory, gives an objective approach to the methodological procedures of the examination of the probability properties of the set of shorter samples and enables derivation of the relationships with the parameters of the whole series sought.

The second and the third sets of problems needed however to be subjected to research. It is evident that the third problem, the solution of which is to yield the values of the design parameters of reservoirs, is the most important.

The computations of the values of the storage function of reservoirs with the help of sets of shorter flow series are closely related to the computation of the adaptiveness of reservoirs to sudden variations of hydrological conditions.

These problems were studied by Patera [90], who emphasized their importance for the optimum design and utilization of reservoirs. The theory of adaptive processes provides plenty of interesting suggestions in this respect. Of particular topicality in this respect is the research into the principles of the theory of adaptive processes as well as its applicability to the dispatcher-type control of reservoirs and water management systems in real time.

The importance of the solution of these problems has recently been proven by Cidlinsky's [24] and Klečka's [47] stimulating studies. They apply the method of the set of shorter series to the solution of the problems of operational control of reservoirs in very short periods, only several years long, immediately linked with the contemporary real hydrological conditions. The papers present the methods of solution and they give concrete examples of the application of these methods to water-engineering practice.

In our research we aimed at clarifying the hitherto unknown properties of the computation of the storage function of reservoirs with the help of extensive sets of shorter flow series and the relationship of these computations to the water-engineering solutions achieved with the help of a single long and stationary flow series.

From the methodological point of view we proceeded so that from the modelled 1000-year series of average monthly flows in selected profiles we gradually generated sets of short random series of length $n = 20, 30, 40, 50$ and 60 years, numbering $v = 10, 20, 50, 100$ and 500 equally long series in one of their sets. The realizations were generated from random series, modelled in two variants, viz. with the input parameters equal to the biassed characteristics of the given real series, and with estimated unbiassed parameters. By analogy with the computations of the design of reservoirs making use of long random series, this method of solution was also applied to the two variants of the sets of chronologically sampled series, and the results obtained were subjected to mutual comparison.

The hydrological computation of the storage function was undertaken using sets of chronologically sampled random series of various lengths n, of extent v, for various combinations of the values of specific storage volumes β and coefficients of minimum-plus runoff α. The result was the ascertainment of the insurance of minimum-plus runoff with respect to repetition p_o, with respect to duration p_t, and with respect to the volume of the water supplied p_d, or the shapes of the regime curves of the type $\beta = f(\alpha)$.

In this case, too, the choice of the combinations of the values of α and β was made with the aim of covering the wide spectrum of the ways of runoff control, viz. from seasonal to markedly long-term. The values of the coefficient of minimum-plus runoff α were chosen, stepped in 0.1 intervals: $\alpha = 0.3, 0.4, 0.5, 0.6, 0.7, 0.8$. The steps of the β values equalled 0.25: $\beta = 0.25, 0.50, 0.75, 1.00, 1.25, 1.50, 1.75$ and 2.00.

The insurance values p_o, p_t and p_d were computed for each chronologically sampled series. The sample values of p_o, p_t and p_d obtained were then further statistically processed for their sets. The average values of all the insurances $\mu(p)$, the maximum values in the set max p, the minimum values min p, and the coefficients of variation and asymmetry were computed from the statistical characteristics available, and quantiles $p_{2.5\%}$ and $p_{97.5\%}$ from the empirical line of transgression of the value of p. For further evaluation no extremes were thus considered at either end of the distribution, viz. twice 2.5%–5% of extremes. And the range of variation of the values (max p − min p), as well as the width of the confidence interval $(p_{2.5\%} - p_{97.5\%})$ were subjected to assessment.

We started by considering the third problem, viz. the design values of the storage volume computed using solely a set of fixed-chronology series. The storage capacity of a reservoir required to ensure the given minimum-plus runoff was computed with the help of both a 1000-year modelled series of average monthly flows and a set of 500 random fifty-year realizations derived from that

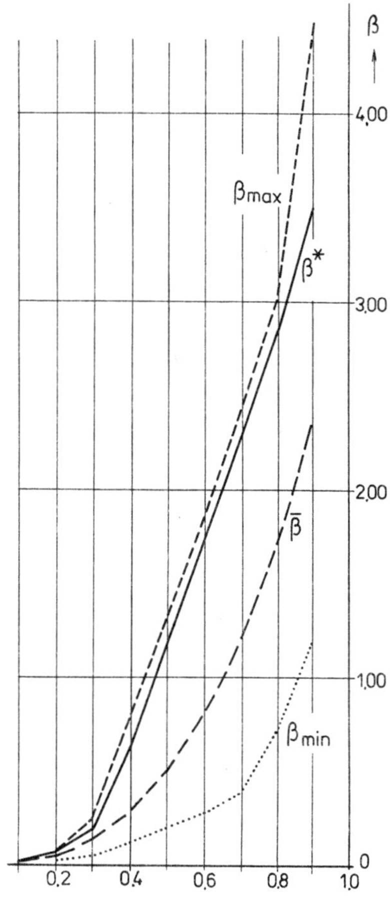

Fig. 57. Regime curves of a reservoir derived from the series of average monthly flows in the Křivoklát profile on the river Berounka: $\bar{\beta}$ – average of values β in their set of 500 random 50-year realizations; β_{max}, β_{min} – marginal values of their set; β^* – regime curve of the reservoir resulting from the solution based upon the original 1 000-year series with unbiassed parameters.

long series. Since the values of the storage capacity obtained using the short flow series have the character of sample values, we assessed their probability properties and compared them with the results arrived at with the help of a long series.

Figure 57 presents an example of the solution, where four curves type $\beta = f(\alpha)$ for a 100% minimum-plus runoff insurance were derived for the Křivoklát profile of the river Berounka, viz. the relationship $\beta^* = f(\alpha)$ using a 1000 year series, the relationship of the expected values of $\bar{\beta}$ of a set of 500 sample values of β, i. e. $\bar{\beta} = f(\alpha)$, and the envelope curves of the maximum and minimum values in that set.

The most interesting are the mutual relationships between all the curves, which yield new knowledge of the behaviour of the sample values of β derived from shorter realizations of a single long flow series. Figure 57 highlights the marked deviation of the curve of the expected values of $\bar{\beta}$ below the values of β^* yielded by the long series. This deviation can be accounted for by the well-known relationship between the results of the water-engineering computation of the storage capacity of the reservoir using a long flow series, and the computation with the help of the sample of that series, which need not of course cover all the periods of water shortage and may thus yield more positive results. The deviation of the expected values of $\bar{\beta}$ is undoubtedly related to the one-sided deviation of the curve of the expected values of the sample coefficients of variation below the coefficient of variation's long-term values.

The result obtained is most significant, for it shows the risk involved in the computation of the storage function of a reservoir using, for example, a set of only several shorter flow series even though they may have been derived from a sufficiently long modelled series. It is obvious that a small set of series like that can yield a result that may on the average be considerably more deviated than that reached with the help of a set of 500 series.

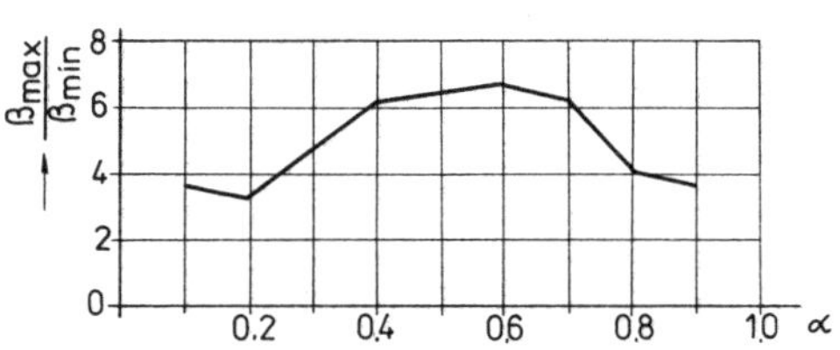

Fig. 58. Ratio of the marginal values of specific storage volumes of the reservoir in the Křivoklát profile on the river Berounka (from Fig. 57) from the solution based upon a set of random realizations for various α.

This danger is well identifiable from Fig. 58, in which is plotted the ratio β_{max}/β_{min} from Fig. 57. Throughout the whole domain of coefficient α the values of β_{max} can be multiples of the values of β_{min}. The design values of the parameters of a reservoir computed from a small set of realizations only can thus be burdened with considerable random errors compared with the computation based upon a long series; and the computation of that type can even result in the respected reservoir being under-dimensioned. From Fig. 57 it can be seen that computing the design parameters of a reservoir from the curve of the expected values of β may be equally risky.

The results obtained led us to abandon examining the confidence interval of the set of 500 values of β at a certain level of significance, as is usual with that type of computation. Such a solution might admittedly reduce the variance range of the maximum and minimum values, but it would not yield any new, revealing information. For this reason, the numerically demanding computations of the storage capacity to ensure the desired minimum-plus runoffs of reservoirs were omitted from the programme of the research. Instead, we studied in more detail the inverse problem, viz. the determination of the runoff insurance based upon the set of realizations for the given volume of the reservoir and the given level of the minimum-plus runoff insurance, as well as their relation to the analogous solution based upon a single long modelled series.

The relatively considerable variance of the values of β for any arbitrarily chosen valie of the coefficient of reservoir-controlled runoff obtained from the

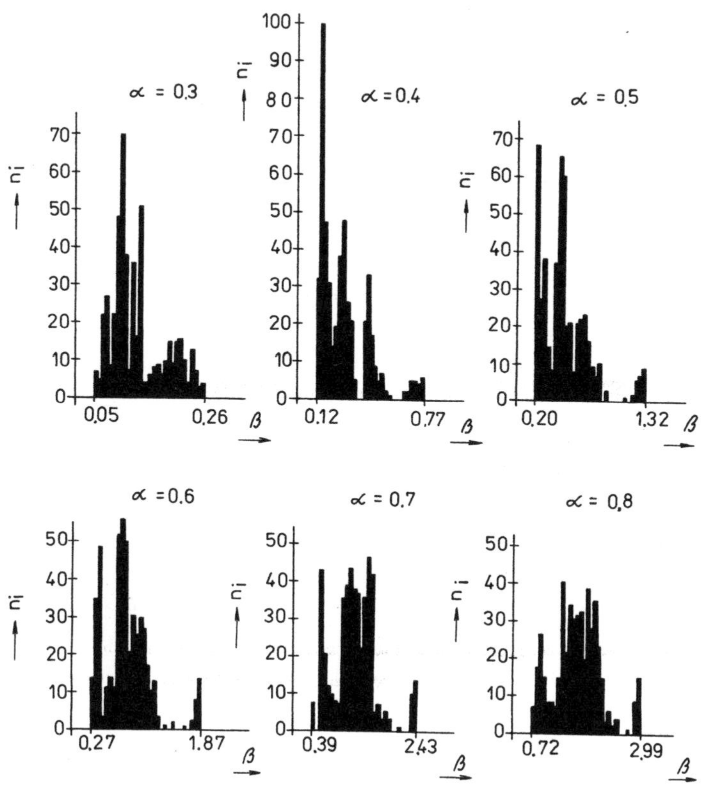

Fig. 59. Empirical histograms of the frequencies of values β from their solution based upon a set of 500 random realizations from the 1 000-year series in the Křivoklát profile on the river Berounka for various values of α.

individual realizations raised a deeper and hitherto unexplained problem of the properties of their probability distribution. We proceeded so that for all the values of α we consistently regarded the corresponding values of β as random variables, and we computed the basic statistical characteristics and constructed the histograms of their 500-element sets.

TABLE 29. Survey of statistical characteristics of 500 values of β derived from 50-year random series of average monthly flows in the Křivoklát profile of the river Berounka

Coefficient of minimum-plus runoff α	Statistical characteristics of 500 sample values of β					
	$\bar{\beta}$	$C_v(\beta)$	$C_s(\beta)$	β_{max}	β_{min}	$\frac{\beta_{max}}{\beta_{min}}$
0.1	0.012	0.312	0.604	0.020	0.006	3.68
0.2	0.049	0.250	0.301	0.075	0.023	3.26
0.3	0.133	0.367	0.761	0.257	0.054	4.81
0.4	0.286	0.513	1.345	0.773	0.124	6.25
0.5	0.503	0.487	1.382	1.318	0.204	6.46
0.6	0.797	0.425	1.077	1.868	0.276	6.77
0.7	1.186	0.356	0.684	2.426	0.386	6.29
0.8	1.675	0.296	0.380	2.992	0.724	4.12
0.9	2.387	0.259	0.385	4.461	1.199	3.73

The results are given in Table 29 and Fig. 59. From the statistical characteristics of the sets of 500 values of β, Table 29 shows interesting courses of the coefficients of variation $C_v(\beta)$ and coefficients of asymmetry $C_s(\beta)$. It is evident now that the β sets exhibit both the highest fluctuation and the highest skewness of distribution for $\alpha = 0.4$, viz. approximately for α at the boundary between seasonal and long-term control. The same facts were published as far back as in 1973 in a study [83] dealing with the design of storage reservoirs based on the random series of average annual flows, and with the derivation of diagrams for the determination of long-term components. The relatively highest fluctuation of the storage capacity at the boundaries between seasonal and long-term cotrols can be accounted for by the fact that in this domain of reservoir-controlled runoff the required magnitudes of β are fairly sensitive to any random variation of the flow regime during a short operating cycle of the reservoir.

The histograms of β display remarkable properties. Their shapes (see Fig. 59) correspond to the calculated values of $C_v(\beta)$ and $C_s(\beta)$, and they manifest a clear tendency, viz. with α increasing, β's fluctuation and skewness becomes less, so that for $\alpha = 0.8$ the histogram assumes a near-symmetrical shape.*) The

*) The histograms differ in scales for β in the individual graphs; the scales were chosen for the $\beta_{max} - \beta_{min}$ range to be divisible into 30 intervals, which were assigned the uniform width of one milimetre in order that the construction of the histograms might be facilitated (before scaling down).

suggestion of a multimode shape of the histograms for all the α's is more difficult to explain; this phenomenon should perhaps be studied closely in a larger number of profiles.

The results of the computation of the design values of the storage volumes of reservoirs arrived at with the help of a set of shorter realizations of flow series and the comparison of these results with the results of the computations based upon a single long modelled series will surely generate considerable interest. It should however above all be clear that there is no question of whether the hydrological computation of the storage function of reservoirs using a set of shorter series is more suitable than the computation based upon a single long series. Each of the two methods is oriented towards the solution of quite a different problem: the set of shorter series serves as a basis for the computation of the storage function of the reservoir in a given shorter period with all the

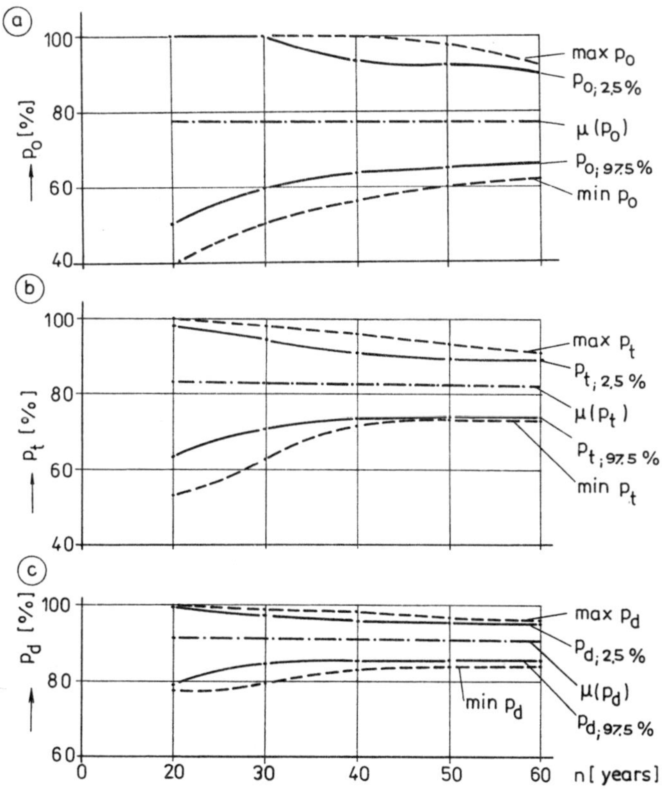

Fig. 60. Effect of the length of a random realization n on the variational range of the values of insurance obtained from the computations of the design of the reservoir in the Křivoklát profile on the river Berounka with the help of 500 realizations ($\beta = 0.25$; $\alpha = 0.8$; a series with biassed parameters).

accidental changes in the flow regime of that period considered, whereas the single long series simulates an often difficult analytic solution in a theoretically infinitely long period.

The computation of the storage function of reservoirs in a certain period for the given α using a set of shorter series leads to a rather wide set of the values of β, from which the expected value of $\bar{\beta}$ should evidently be selected as the design value in accordance with the principles of the theory of probability. This methodological procedure is the more tempting, because it can help to cut investment costs of reservoir design. The solution is however biassed and it involves an extraordinary risk of the reservoir failing under adverse hydrological conditions outside the design-covered period.

It is therefore more reliable to base reservoir design upon a single long modelled series with estimated unbiassed parameters.

Far more complex however is the problem of operating completed storage reservoirs under changing hydrological conditions. That is why we attempted the solution of the inverse problem, viz. computing the variation of runoff insurance with respect to repetition p_o, duration p_t, and the volume of the water supplied p_d, using various realizations of the flow series for the given volume of the reservoir β and the given coefficient of minimum-plus runoff α [88].

Figures 60 and 61 visualize the dependence of all the three indicators of runoff insurance upon the length of 500 realizations derived from a 1000-year modelled series of average monthly flows in the Křivoklát profile of the river Berounka. Figure 60 presents the solution with the help of a random series with biassed (uncorrected) parameters and with $\beta = 0.25$ and $\alpha = 0.8$. Figure 61 then shows the result of using a series with unbiassed parameters and with $\beta = 1.00$ and $\alpha = 0.8$.

The two examples coincide in showing that even with a markedly different relative magnitude of the storage volume β, the runoff insurance fluctuates considerably in the invidual realizations of the flow series. This is a satisfactory proof of the complexity of runoff control under changing hydrological conditions. The character of the fluctuation of the runoff insurance does not change very much even if the extreme values, max p and min p, are eliminated and substituted by critical values of the level of significance equal to 5% (viz. the confidence interval thus closing with values $p_{97.5\%}$ and $p_{2.5\%}$ always for the respective p_o, p_t and p_d).

Another interesting property of the curve of runoff insurance is the dependence of the fluctuation of that insurance upon the length of the realizations of the flow series n. It turns out that the bias of both the extreme and the critical values diminishes with n increasing. These relationships correspond to the basic properties of the bias of the statistical characteristics of the individual realizations, as well as the convergence of their expected values towards long-term parameters.

It can be shown that a lower fluctuation of the hydrological regime can have a positive effect upon the fluctuation of the runoff insurance in the individual realizations; on the other hand, a high fluctuation involves the danger of considerable fluctuation of the runoff insurance in shorter realizations.

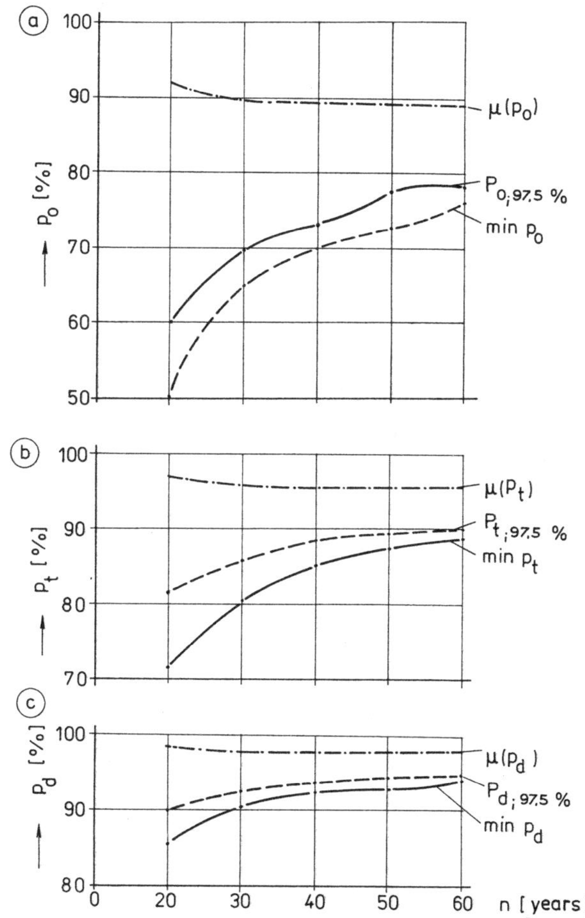

Fig. 61. Effect of the length of a random realization n on the variational range of the values of insurance obtained from the computations of the design of the reservoir in the Křivoklát profile on the river Berounka with the help of 500 realizations ($\beta = 1.00$; $\alpha = 0.8$; a series with unbiassed parameters).

This fact is of particular importance as far as the control of the operation of reservoirs under hydrological regimes with higher variability of runoff is concerned. Under these conditions timely adjustments of the directions for operating water-engineering structures will necessarily have to be reckoned with.

Noteworthy relationships arise between the expected values of the 500-term sets of runoff insurance $\mu(p)$ and long-term insurance p^* calculated for the whole 1000-year series. For the sake of greater clarity, the extreme and the critical values of the sample insurances have been omitted in Figures 62 and 63, a comparison has however been made of the pairs of curves $\mu(p)$ and p^* for the random series modelled with both biassed and unbiassed parameters. The solution has produced the following results:

a) the two curves $\mu(p)$ and p^* derived from a random series with unbiassed parameters gives invariably higher runoff insurance than the same pair of lines

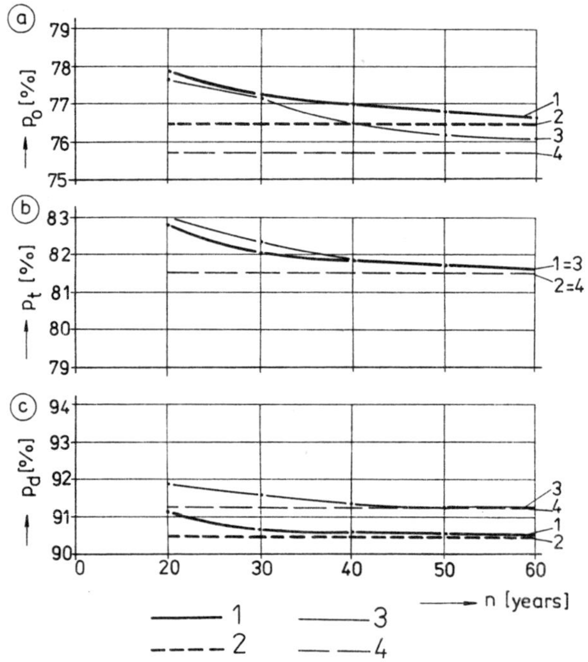

Fig. 62. Effect of the length of a random realization n on the expected value of insurance obtained from the computation of the design of the reservoir in the Křivoklát profile on the river Berounka involving 500 realizations ($\beta = 0.25$; $\alpha = 0.8$):

Curve *1* – $\mu(p)$ from the solution using a set of random realizations of a series with biassed parameters;

2 – p^* from the solution using a long series with biassed parameters;

3 – $\mu(p)$ from the solution using a set of random realizations of a series with unbiassed parameters;

4 – p^* from the solution using a long series with unbiassed parameters.

derived from a random series with biassed parameters. We dealt with this mutual relationship in detail when we were computing the long-term stationary function of reservoirs, from which it followed that the compensation of the statistical characteristics of a real series with systematic errors, and the introduction of these characteristics into the mathematical model, could result in more positive results of the hydrological solution concerning the storage function of reservoirs;

b) in both the variants of random series, with the length of the realizations, n, of the series increasing, the curves of the expected values of runoff insurance $\mu(p)$ approximate to long-term values of that insurance p^*. Thus, as long as the insurances found with the help of the individual realizations are regarded as random variables, the convergence of the $\mu(p)$ curves mentioned above will indicate the consistency of the estimates of runoff insurance in the same way as in the case of the consistency of the estimates of the statistical parameters of flow series;

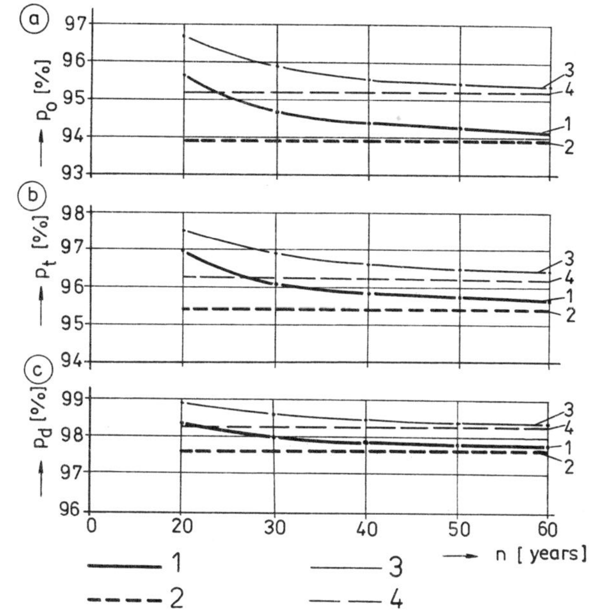

Fig. 63. Effect of the length of random realization n on the expected value of insurance from the computation of the design of the reservoir in the Křivoklát profile on the river Berounka ($\beta = 1.00$; $\alpha = 0.8$). (For curves *1, 2, 3, 4* see the text to Fig. 62).

c) undoubtedly the most interesting is the one-sided deviation of the curves above the respective long-term values of p^*. This relationship is inverse to the analogous biassed relationships between the expected values of the sample coefficients of variation and asymmetry and their corresponding long-term parameters. The relationship between $\mu(p)$ and p^* is however quite logical. As can be seen from Fig. 57, the hydrological computations on the basis of shorter realizations can yield more positive results (viz. smaller storage volumes, higher runoff insurance for a given volume etc.) than in the case of a single long random series including the most favourable drier periods. The solution of the inverse problem, viz. the problem of finding the runoff insurance p for the given volume β and the given coefficient of minimum-plus runoff α on the basis of the individual realizations, will thus also yield the expceted values $\mu(p)$ slightly higher.

The results of the computation of the fluctuation of runoff insurance arrived at with the help of short realizations and the fluctuation of the required storage volumes ensuring minimum-plus runoff are thus in full agreement. The results show that the computations concerning the operation of reservoirs on the basis of shorter realizations of the flow series require some prudence – the same prudence as when β is to be suggested for a given α. In this case we are tempted by the one-sided deviation of $\mu(p)$ above p^* to raise the minimum-plus runoff, which may however result in a failure of the reservoir under adverse hydrological conditions. A single long modelled series is thus a far more reliable basis for the computations concerning the operation of reservoirs than a set of shorter series.

Despite the risks involved in the method of computing the storage function of reservoirs on the basis of a set of shorter realizations of the flow series, this method can nevertheless prove useful as a supplementary means of computing the storage function of the reservoirs already completed under various hydrological conditions, particularly a means of dealing with the probable fluctuations of the minimum-plus runoff insurance, or the fluctuations of the minimum-plus runoff with a given volume and degree of insurance.

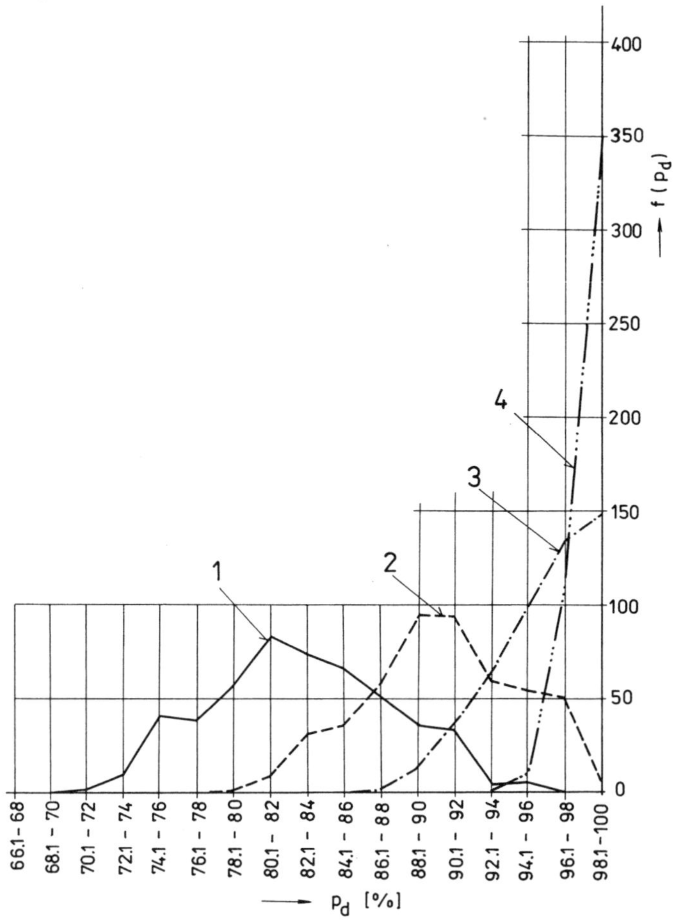

Fig. 64. Empirical distributions of the frequencies of the values of insurance p_d in the Křivoklát profile on the river Berounka in a set of 500 forty-year random realizations ($\beta = 0.25$; 1 000-year series with unbiassed parameters): *1* – $\alpha = 0.8$, *2* – $\alpha = 0.7$, *3* – $\alpha = 0.6$, *4* – $\alpha = 0.5$.

Besides, it should be evident that the phenomenon exposed, viz. the fluctuation of runoff insurance with shorter realizations, will have to be taken into consideration in further elaboration of the methods of hydrological computations concerning reservoirs, particularly in the construction of stochastic models of the control of their operation under extreme conditions.

Research into the variable properties of runoff insurance computed from a set of shorter realizations of the flow series also covered the problem of the dependence of the runoff insurance upon the length of the series n, and upon their number ν [88]. It turned out that with a particularly low number of very short realizations the solution was burdened with the greatest random errors. This result prompts us to use considerable caution in designing reservoirs, because for reasons of economy water-engineering practice very often resorts to several short modelled series (e. g. ten 50-year series) as a satisfactory hydrological basis. Even if the expected values of the set of the results based upon shorter series are accepted as design parameters of a reservoir, a solution of this type can be linked with a high risk of the reservoir being under-dimensioned. A similar risk is involved in the computation of the parameters of the ancillary systems of reservoirs if that computation is thus unsatisfactorily based.

Interesting probability properties are exhibited by the histograms of runoff insurances with respect to the volumes of the water supplied, p_d, ascertained with the help of a set of shorter realizations of the flow series. Figure 64 shows plots of the empirical distribution of these runoff insurances for the selected $\beta = 0.25$ and various α's in the Křivoklát profile of the river Berounka. The curves of the individual α's are remotely reminiscent of the curves of distribution χ^2 (chi-square distribution) for various degrees of freedom (the role of the degree of freedom resting with α). With an increasing number of degrees of freedom these curves approximate to the curve of Gaussian distribution of probability. Proving this similarity is however in no way so easy. But even without such a proof, the mutual relationships of the curves plotted in Fig. 64 will certainly engender interest, for they are highly informative as far as the character of the regularities of the fluctuations of the minimum-plus runoff insurances based upon shorter realizations of the flow series are concerned.

12.3 Effect of the estimation of the autocorrelation function of flow series on the computation of the design parameters of storage reservoirs

Designing storage reservoirs involves above all the determination of their long-term component, the magnitude of which depends upon a reliable estimate of the probability properties of the series of average annual inflows feeding the reservoir, including their autocorrelation function. The second major problem is the transformation of the afflux, viz. the control of the afflux. The solution to the problem is thus easier to reach, the simpler the mathematical model of the afflux and the transformation of this model. And that is also why the original models derived several years ago and linked with the names of Hazen [36], Sudler [105], Kritskiĭ and Menkel' [58], Savarenskiĭ [101] and others, were based upon the simplest assumption, viz. the influx into the reservoir was looked upon as an absolutely random series of discrete variables representing the annual volumes of runoff or average annual flows, and runoff was considered constant.

The following development of the solution to this problem however showed that the assumption of the independence of the annual inflows feeding the reservoirs was in no way always fully justified, because in the flow series certain autocorrelation tendencies were discovered influencing the required storage volume of the reservoir. In 1936, P. A. Efimovich [116] pointed to this fact when he analyzed twenty-four rivers of the European part of the ex-Soviet Union. He formulated the relationship between the coefficient of variation of the annual modules of runoff and the coefficients of correlation between the runoffs in both the neighbouring and the more remote years.

The assessment of the internal relationships of correlation in hydrological series, which can generally be expressed by autocorrelation functions, gives rise to two major problems. The first is the physical essence of the relationships of correlation and the possibility of assessing the representativeness of these relationships with respect to future regimes, which is of the utmost interest. Some authors explain the stochastic dependence of the annual flows by transfer of the volumes of the water stored from year to year, but it seems that this view may be of rather limited validity, for in several cases autocorrelative relationships between the annual flows were ascertained where the river was drying up.

Hydrological experts have recently been endorsing the opinion that the autocorrelative relationships must be studied in a larger number of profiles of more extensive territorial units. We are of the opinion that such an approach can be most effective provided that genetic aspects are taken due account of in statistical analyses.

The explanation of the physical meaning of the autocorrelation function of hydrological series was attempted by Yevjevich [120] in the years 1963 and 1964. He undertook extensive statistical analyses of thirty seven 150 year flow series obtained from 140 stations on the rivers of the whole world (viz. 72 stations in the United States of America, 13 Canadian, 37 European, 11 Australian, 4 African and 3 Asian stations). He also made an assessment of 446 flow series of North West America and 141 precipitation series of the same region. On the basis of the rich material collected, he infers that the sequences of hydrological quantities will hardly provide a proof of a cyclic (or deterministic) trend, and he concludes that the sequence of wet and dry years should be regarded as absolutely random.

The periodic properties of the hydrological regimes in the Czechoslovakia were studied by Vitha [112, 113] Souček [102, 103] and Souček and Vitha [104]. The results of their research are most valuable, for their assessment covers the genesis of the rivers, and the dependence of the moving statistical characteristics of the longer flow series upon time is substantiated by similar properties of other hydrological and meteorological series (e. g. precipitation, air temperature, sun-spots etc.) and their correlation. And Balek and Anděl's [5, 7] research also showed that the sequences of hydrological variables can occasionally manifest autocorrelative tendencies and periodic components.

Nacházel [71] discovered certain autocorrelative tendencies – mostly of a harmonic type – in some longer annual flow series. Autocorrelation functions were analyzed for selected Czechoslovak and other rivers with longer rate-of-flow observations.

Svanidze's book [108] also sets out to find autocorrelative tendencies in flow series.

The long-term investigation of autocorrelative tendencies of the annual flow series carried out so far has exposed the considerable complexity of the problem. It can hardly be doubted that the autocorrelative tendencies of these series can be fully demonstrated and that the periodic character can be revealed. Since their statistical significance and their genesis will however be more difficult to prove, we are of the opinion that these problems require further study.

The assessment of the effect of the autocorrelation function of the annual flow series on the magnitudes of the long-term components of reservoirs is also a relatively difficult task. The analytic solution of this problem is rather complex, as shown by Moran's original approach [69] to the problem of the distribution of the probability of a reservoir being stored to capacity on the boundaries of a time interval. Even in the elementary case of absolutely random affluence, the probability density of the sum of the affluence and the capacity stores will lead to the computation of extremely complex integrals. Discretization of the random variables and approximation with the help of a system of linear equations is therefore inevitable.

From the many modifications of the original Moran method (e. g. Moran [70], Gould [33], White [118]), the Lloyd modification (from the year 1963) is the most significant from the point of view of long-term runoff control. This modification replaces the absolutely random sequence of flows by a simple Markov chain [66] and is thus well applicable to the computation of seasonal fluctuations of affluence and delivery. The assumption of a simple Markov chain may however prove to be a limiting condition, which may not always be fully satisfied.

Another method of deriving the long-term components of a stored volume was applied by Kritskiĭ and Menkeľ [58], originally under the assumption of absolutely random affluence into a reservoir. Later on, these authors modified the method and suggested deriving the long-term components under the assumption of random affluence into the reservoir in the form of a simple Markov chain [60]. For the first method Pleshkov worked out diagrams facilitating the determination of the long-term components given the parameters of Pearson's IIIrd type distribution, the required reservoir-ensured delivery α, and runoff insurance according to repetition p_0. Similar diagrams were designed by Guglij for the second Kritskiĭ–Menkeľ method, but also taking account of the coefficient of correlation of the flows in the neighbouring years $r(1) = 0.30$.

Storage reservoirs are considerably easier and more flexible to deal with when synthetic flow series are used. The advantage of such approach, as compared with the preceding method, is above all the fact that the storage functions of reservoirs can be dealt with with the help of synthetic flow series with any type of probability distribution and autocorrelation function. And it is by no means a negligible numerical advantage that in the computations use can be made of a simple balancing method known from the computations with the help of real flow series. These algorithms are thus easily programmable and controllable.

The simplest ways of forming synthetic hydrological series were applied by Hazen (1914), then Sudler (1927) and Ivanov (1946). The drawback of these methods correspond to the general level of the methods of mathematical statistics in those times, and they also reflect the level of knowledge of the complex laws of hydrological regimes under various geographical conditions, as well as the lack of modern computer technology.

The mathematical models of hydrological processes had not been studied more systematically until the early sixties. In the development of engineering hydrology and the theory of reservoir-controlled runoff the application of the apparatus of the autoregression models to the generation of synthetic series was of extraordinary importance. Numerical modelling of the processes of river runoffs by Monte Carlo methods was first considered for correlative relationships only between the neighbouring terms of a series, which were then expressed with the help of autoregression of the first order. These methods of modelling synthetic series were published by Svanidze [106] in 1961. Analogous methodological procedures were published by Fiering [32] in the same year.

The methods of modelling synthetic series were later elaborated in more detail and generalized by the introduction of the assumption of the compound Markov chain, which began to be applied to the more complex autocorrelation structures of hydrological series (Svanidze [107]). As we have already shown in Chapter 7, the broad Box–Jenkins' methodology was of equal importance for the modelling of synthetic series. For the historical development of the modelling of the time series in hydrology the reader is referred to the [100] monograph.

A sufficiently powerful computer is practically indispensable in this respect. And the same applies to hydrological computations concerning water reservoirs. Where a computer is not available, the determination of the required magnitudes of the long-term components of the storage volume is facilitated by diagrams. In his work mentioned above ([107]) Svanidze published such diagrams enabling determination of the long-term components of reservoirs, elaborated on the basis of synthetic series of average annual flows. These diagrams are more extensive than Pleshkov's and Guglij's original diagrams: they go as far as the value of the coefficient of correlation of the flows of the neighbouring years $r(1) = 0.6$. As far as their quality is concerned, they are however based on the same assumptions as the preceding Kritskiĭ–Menkel's method, viz. that the probability distribution of the affluence into the reservoir follows the Pearson IIIrd type curve, and the correlative relationships are considered only between the affluxes of the neighbouring years.

Reznikovskiĭ [96], too, pays great attention to the modelling of hydrological series, particularly as far as the application to the problems of hydrological energy generation is concerned. Reznikovskiĭ's work, too, contains new diagrams facilitating the determination of the long-term components of storage reservoirs. In correlation analysis, the same assumptions were built on as in Svanidze's [107] diagrams; the applicability of Reznikovskiĭ's diagrams is however further extended, viz. they reach as far as the value of $C_v = 1.5$, and for $C_s = 1.5\ C_v$ and $C_s = 4C_v$ an attempt is made at mitigating the drawbacks of the Pearson distribution (for $C_s \neq 2C_v$) by making use of the triparametric gamma distribution.

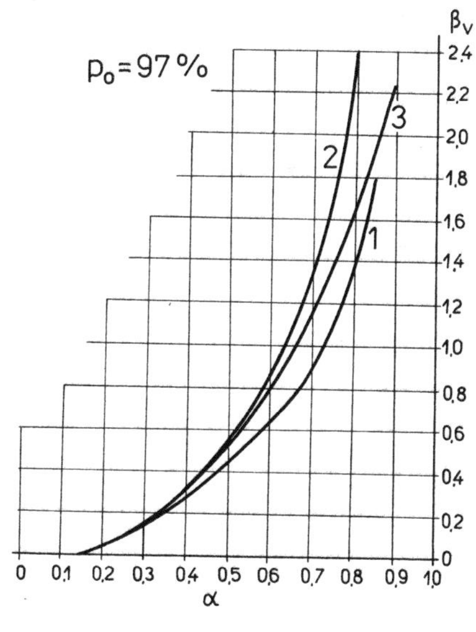

Fig. 65. Relationships $\beta = f(\alpha, p_0)$ for an absolutely random sequence *1*, simple Markov chain *2* and composite Markov chain *3* $C_v = 0.5$, $C_s = C_v$, $r(1) = 0.3$.

From all these works a rather large effect of the coefficient of correlation $r(1)$ on the required size of the storage volume of the reservoir can be safely assumed. Figure 65 presents and example of the determination of the long-term component of the storage volume β_v under the assumption of an absolutely random sequence of the affluences into the reservoir and also the assumption of the flow series with $r(1) = 0.3$. It turns out that the effect of $r(1)$ on the desired magnitudes of the long-term components β_V becomes more intensive with the volume of the reservoir-supplied minimum-plus runoff α increasing. It can similarly be shown that the effect of $r(1)$ on β_V increases with the value of the coefficient of flow variation C_V and runoff insurance p_o rising. A dependable estimation of the value of $r(1)$ is thus of immense importance as far as the design parameters and investment costs of water works are concerned.

For this reason we were particularly interested to know the effect of the other ordinates $r(k)$ of the autocorrelation function besides the coefficient of correlation $r(1)$, especially in cases where this function is of a harmonic character, which is relatively very frequent in the real flow series. Finding a generally valid theoretical autocorrelation function of this type is however an extremely difficult task and one that is practically insolvable, for the modelled structures of the flow series are rather varied and referrable to a number of factors shaping the runoff process. Under these circumstances, the "design" curve of the autocorrelation function was considered in the shape of damped harmonic motion, the initial course of which, as well as its periodic properties, corresponded quite well to some of the empirical autocorrelation functions found in the Elbe and some other river basins. These empirical autocorrelation functions were approximated to by a theoretical curve of the following form:

$$r(k) = \frac{5}{3} r(1)\, e^{-0.1k} \,.\, \cos \frac{2\pi}{15}(k + 1) \quad \text{for} \quad k \geqq 1 \,,$$

$$r(k) = 1 \qquad\qquad \text{for} \quad k = 0 \,. \tag{12.1}$$

The calculated values of this curve for the $r(1) = 0.30$ chosen are given in Table 30; its graphical representation is shown in Fig. 66. The properties of this

TABLE 30. Ordinates of damped harmonic autocorrelation function

k	$r(k)$	k	$r(k)$	k	$r(k)$
0	1.000	11	0.052	21	−0.060
1	0.303	12	0.101	22	−0.055
2	0.127	13	0.125	23	−0.042
3	−0.039	14	0.124	24	−0.023
4	−0.168	15	0.102	25	−0.005
5	−0.245	16	0.068	26	0.012
6	−0.269	17	0.029	27	0.023
7	−0.243	18	−0.009	28	0.028
8	−0.182	19	−0.038	29	0.028
9	−0.102	20	−0.055	30	0.023
10	−0.020				

curve correspond well to the empirical autocorrelation functions: in a number of cases we could demonstrate their period of approximately 13–15 years; the first negative extreme very often occurs in the region round $r(6)$ and $r(7)$ with

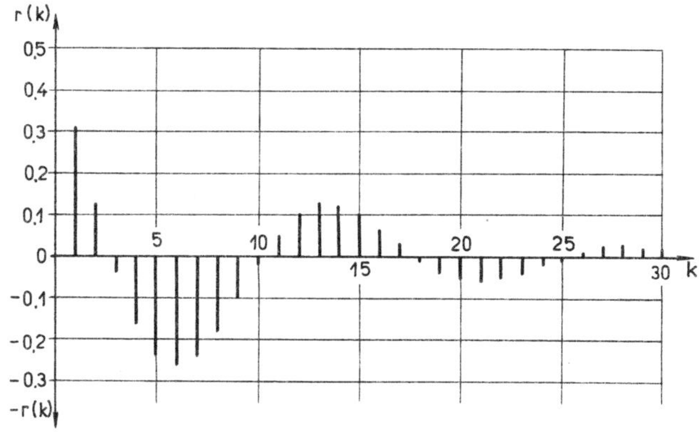

Fig. 66. Ordinates of the damped harmonic autocorrelation function.

the absolute value not differing significantly from the $r(1)$ value. The approximation also revealed another significant property of some of the empirical autocorrelation functions, viz. a relatively fast fall-off of the initial positive values and their conversion into negative ones.

As the next phase of the solution, we modelled synthetic 1000-year autoregressive sequences for function (12.1), i. e. invariably for the Pearson IIIrd type distribution with the following parameters: the expected values were selected as unities, the coefficients of variation C_v in eight alternatives with the values of $C_v = 0.10, 0.20, \ldots, 0.80$, the coefficients of asymmetry for each value of C_v in three alternatives, $C_s = C_v$, $C_s = 2C_v$ and $C_s = 3C_v$.

In view of the fact that the output parameters of every modelled synthetic sequence are biassed by certain random errors, we increased the dependability of the solution by modelling for each combination of input parameters ten synthetic sequences, which were then further statistically evaluated. The total number thus equalled 240 synthetic sequences, which were then used for the computation of the long-term components of the volumes of reservoirs, viz. for the runoff insurances according to recurrence $p_o = 90\ \%, 95\ \%, 97\ \%$ and $99\ \%$.

For the significance of the solution the reader is referred to Fig. 65. If, for the autoregressive sequence of the first order, the long-term components of the volume of a reservoir increase markedly with the reservoir-supplied minimum-plus runoff α rising, as compared with the absolutely random sequence (curves *2* and *1*), then the long-term components of the reservoir volume computed using an autoregressive sequence of a higher order continuously lose the shape of curve *2* and assume the course of curve *1*. The required magnitudes of the long-term components are thus relatively lower (as compared with the analogous solution with the help of the autoregressive sequence of the first order), so that the introduction of a more fitting autoregression function into the model of an autoregressive sequence of a higher order can well lead to a more economical design of reservoirs.

The mutual relationship of curves *1*, *2* and *3* visualized in Fig. 65 can be explained above all by the effect of the individual ordinates of the autocorrelation functions on the internal structure of the synthetic sequence modelled (on the rise of tendencies in the chronological arrangement of the elements of the

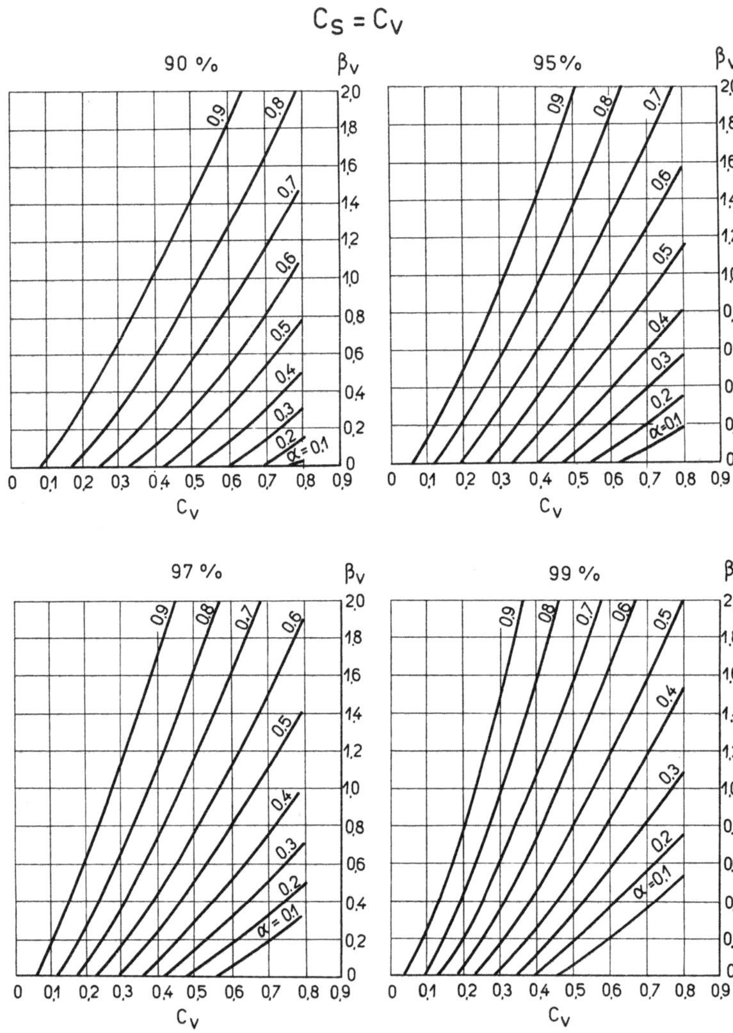

Fig. 67. Diagrams for the determination of long-term components β_v of the storage volume of a reservoir; $C_s = C_v$.

sequence), thus also on the required magnitudes of the long-term components of the volume of the reservoir. Curve *2*, as compared with curve *1*, is an expression of higher requirements concerning the magnitudes of the long-term components due to the fact that the autoregressive sequence of the first order manifests adverse tendencies to aggregate the dry years given by the positive ordinates of the autocorrelation function $r(k) = r(1)^k$. Curve *3*, in contrast, leads to more economical designs of reservoirs, because the autoregressive sequence of a higher order also manifests, apart from the adverse tendencies to aggregate the dry years, the more positive tendencies to aggregate the wet years

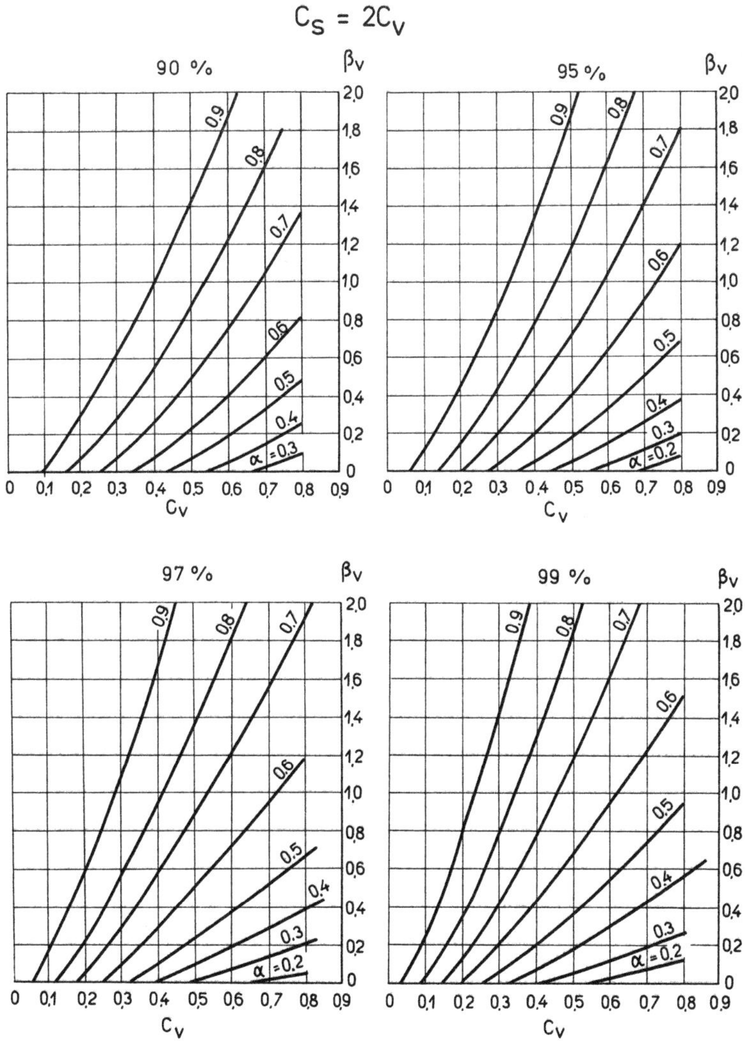

Fig. 68. Diagrams for the determination of long-term components β_v of the storage volume of a reservoir; $C_s = 2C_v$.

given by the negative ordinates of the autocorrelation function. These two tendencies can offset each other, so that the total effect of the harmonic correlation function can manifest itself in the design of a reservoir approximately as a fictitious case of an absolutely random sequence of annual flows (with curve *3* approximating to curve *1*).

In the last phase of the computation, diagrams were constructed from all the calculated values of the long-term components β_v for all the combinations of the input parameters of synthetic sequences, giving expression to the relationship $\beta_v = f(C_v, C_s, \alpha, p_o)$. These diagrams are presented in Figures 67, 68 and 69. Their

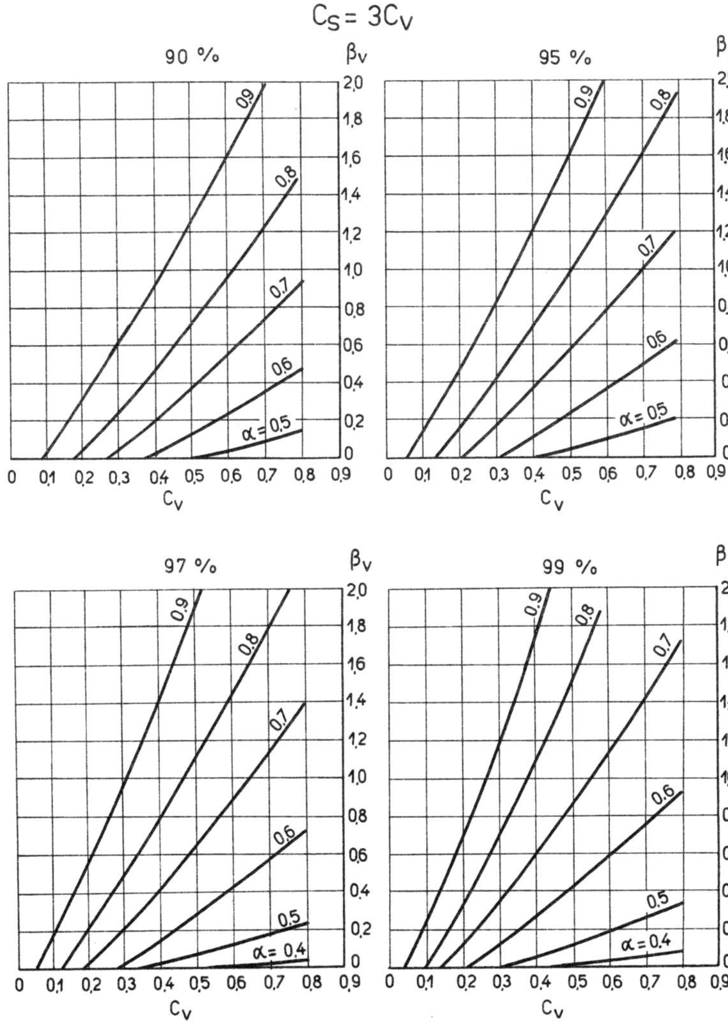

Fig. 69. Diagrams for the determination of long-term components β_v of the storage volume of a reservoirs; $C_s = 3C_v$.

application is very simple: for the given input statistical parameters, the expected delivery α and the insurance of that delivery, the required magnitude of β_v can easily be read, and the inverse problem of finding α for the given β_v is equally simple. The diagrams also enable easy interpolation.

The solution suggested above shows very well the effect of the shape of the curve of the autocorrelation function of the average annual flow series upon the required sizes of storage reservoirs as well as upon the possibility of their more economical design by more suitable autocorrelation functions being introduced into the respective mathematical models of the synthetic sequences. The solution also shows considerable prospective importance of the application of the synthetic sequences as a hydrological basis for designing reservoirs and hydrological systems, and optimization. In cases where analytical methods fail to solve these problems, or where these methods have not been derived at all owing to the difficulty of such an undertaking, the solution can be attempted with the help of satisfactorily long synthetic flow series which can approximate to the exact analytic solution with a high degree of reliability.

Another consideration offers itself in this context. It is obvious that for the construction of fitting mathematical models of synthetic sequences, the analysis of both the autocorrelative properties and parameters, and the type of the probability distribution, including the possible non-stationary tendencies, are of great importance. The analysis of these properties, as well as their estimation, therefore remains one of the current tasks; this brings us back to the importance of the problems of time series that has already been discussed in Chapter 7.

12.4 Relationship between estimation theory and optimum control of reservoirs in real time

12.4.1 Basic problems of optimum control of reservoirs in real time

Optimum usage and control of reservoirs in real time is one of the most important fields of solution of the problems of water resource systems. And this field is fairly complex and extensive, for it comprises a large number of theoretical and practical water-engineering problems. The basic problem is the fact that control of reservoirs and their systems is effected under conditions of stochastic uncertainty: the development of hydrological conditions, and very often also the development of the requirements concerning the usage of reservoirs, are not precisely known in advance. Unlike the process of designing reservoirs, where design hydrological situations are invariably worked with, the control of reservoirs needs to be dealt with under any situation, the development of which will of course be completely unknown in advance.

The uncertainty of the development of the hydrological conditions can to a certain extent be reduced by short-term stochastic forecasts; but the success of these forecasts and their period of head start are however conditional upon a large number of factors. Under actual conditions the utilization of a forecast can also sometimes be greatly limited by practical circumstances. For some river basins, for instance, no reliable precipitation and runoff models may be derived and the interests of some operators of energy-generating equipment may be contrary to the requirement of timely drainage of reservoirs before flooding periods.

Another difficult problem is posed by the multipurpose utilization of reservoirs and the multivariate optimization of control linked with it. Some of the purposes for which a reservoir has been constructed may even be mutually incompatible and some of the criteria followed may defy quantification, so use must be made of methods enabling aggregation of both the qualitative and the quantitative criteria in order that the resulting assessment may be as complete as possible.

The control of a reservoir in real time will have to take account of a large amount of information on the state of the river basin, which must be processed by efficient computers in the control stations of the basin. For research purposes it therefore means deriving decision models that would use that information to transform it into a base capable of coping with any operating situations.

Water-engineering practice, on the other hand, expects the decision models to be as simple as possible, so that in case of need they may be readily derived even with the help of the less powerful personal computers. Viewing the problem from this aspect we can see considerable gaps between systems theories and practice in the field of reservoir control. It is therefore essential that reliable assessment should be made of the possibilities of concrete application of the systems approach and the optimization techniques in the solution of practical problems.

The selection of the method of control is of fundamental importance in this respect. It depends upon a number of factors, the most important of which are, for instance, the type and the structure of the water resource system, the aims and the criteria of the control, and the hydrological and other information available in the given case for the derivation of the decision model. Experience shows that it is far from expedient to attempt to derive a single model suiting all the purposes of a water resource system. A set of partial models (for instance, a flood control model, a dry-year control model etc.), which can be readily applied in various operating situations to get an immediate effect will certainly prove much more helpful.

A number of interesting methods have been recently devised in the systems-oriented literature of designing reservoirs and entire water resource systems, which have of course also been subjected to critical appraisal [97, 119]. These studies are valuable not only because they present the achievements and experience gained in that field of knowledge, but also because they set out the still open problems that are to be solved.

12.4.2 Possibility of applying the principle of adaptivity to the control of reservoirs in real time

In our research we were concerned with a set of simple decision models for the control of the storage function of reservoirs in dry periods. The models make full use of the fundamental principles of the theory of adaptive processes and short-term stochastic forecasts of the afflux into a reservoir. The effort to derive simple decision models was motivated primarily by practical needs.

The basic problem of the derivating of a decision model for the control of reservoirs in real time is the algorithmization of the operations in a dry period. In our case the algorithm of the operations was related to the actual volume of the water stored and the expected development of the hydrological situation. The control of reservoirs thus adjusts itself continuously to the actual values of these parameters in harmony with the principle of adaptivity. The decision model is however conceived of as an open system, the controlling function of which can also be adjusted in order to suit further requirements, viz. the parameters of its environment.

The decision model was formally expressed as a matrix, the elements of which stand for the limited supplies of water in a dry period, related to the two parameters mentioned above. An example of this kind of matrix is shown in Table 31, from which it follows that in dry periods, limited deliveries from the

TABLE 31. Example of decision model R

$V_d/V_{disp,d}$ \ Q^{pr}/Q_n	$\geqq 1$	0.99–0.80	0.79–0.60	0.59–0.40	<0.40
1	1	1	1	1	1
0.51–0.99	1	1	1	1	1
0.26–0.50	1	0.95	0.90	0.85	0.80
0.01–0.25	1	0.90	0.80	0.70	0.60
0	k_p	k_p	k_p	k_p	k_p

Notes:

1. k_p – module coefficient of natural influx into reservoir Q_p that can be drained off with the reservoir empty $k_p = Q_p/Q_n$;
2. the other coefficients in the table express the relationship between the controlled discharge from the reservoir on day $d + 1$ and the required long-term guaranteed runoff, viz. Q_{d+1}/Q_n;
3. V_d – actual filling of the storage volume of the reservoir on day d,
   $V_{disp,d}$ – the dispatching operator's filling up of the storage volume of the reservoir on day d, i.e. the required filling up of the reservoir in order that long-term discharge Q_n may be ensured as desired.

reservoir are started as soon as the level in the reservoir manifests a tendency to fall. The volume of the water saved in this way can then be effectively used in a critical situation when the reservoir is empty and when the highest economic losses due to the non-delivery of water are to be expected.

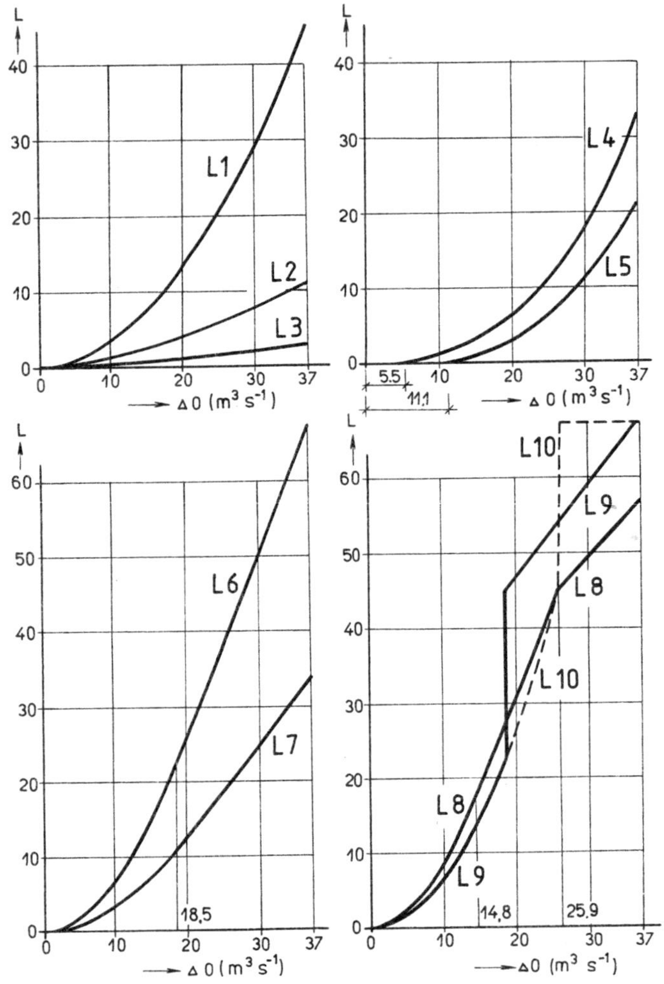

Fig. 70. Examples of loss functions for estimating economic losses caused by limited water delivery.

The minimum economic losses due to limited delivery of water in dry periods were regarded as the criterion of optimality. Since under the conditions of stochastic uncertainty the character of these periods remains of course unknown, the minimum economic losses were computed for all the dry periods of the long synthetic series of average daily flows. The losses caused by the limited deliveries of water were ascertained for ten various types of hypothetical loss functions giving expression to the relationship between the volume of the limited deliveries of water and the daily losses (Fig. 70).

The aim of our research was to find a decision model that would guarantee minimum economic losses throughout the whole flow series for all the types of loss functions considered. This solution also made it possible for us to assess the

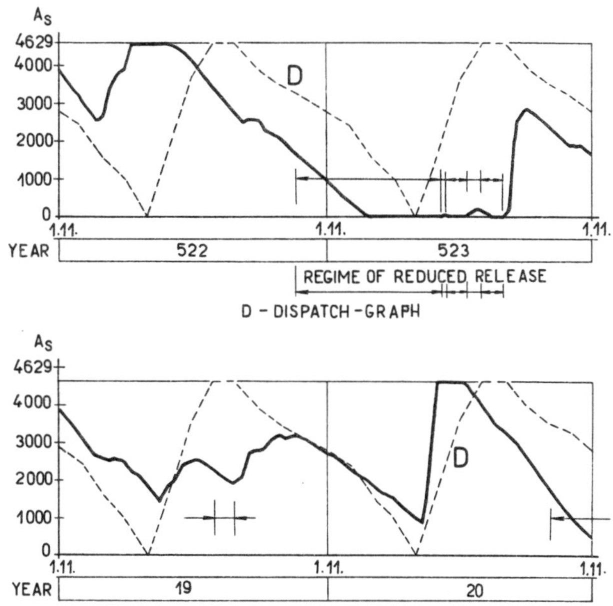

Fig. 71. Simulated curves of the filling and the emptying of the Orlík reservoir on the river Vltava (Czechoslovakia) plotted on the basis of the modelled series of average daily flows.

sensitivity of adaptive control to the individual types of loss functions, as well as to judge the adequacy of that control as compared with non-adaptive control.

The basic problem of making decisions under stochastic indeterminacy is visualized in Fig. 71, presenting two main characteristic simulated examples of the course of the filling and emptying of the Orlík reservoir on the river Vltava in dry periods. The upper part shows a critical dry period ending in the reservoir being completely emptied and the ensuing hydrological failure in the delivery of water. The economic losses can in this case be alleviated by preventive limitation of deliveries imposed as soon as the reservoir starts emptying.

The bottom part of Fig. 71 is a representation of a failure-free dry period that emptied half the reservoir only. The possible pessimistic approach of a controller under the conditions of indeterminacy can in this case lead to the limitation of water deliveries, which can however cause unnecessary economic losses*). This will of course weaken the effect gained by adaptive control in the critically dry periods with the reservoir completely emptied. It is obvious that under the conditions of indeterminacy decision-making in real time has the character of an

*) In that type dry period there would be no such losses provided the reservoir was used so as to guarantee long-term delivery, viz. free from any adaptation.

optimization problem, which should be solved simultaneously for all the dry periods.

From what has been said above it follows that the adaptive model operates with a certain loss, which is given by the probability character of the hydrological conditions, viz. by their indeterminacy. Stochastic adaptation can thus only approximate to optimum control. (In a theoretical limit case, optimum control can be achieved only on the basis of complete a priori deterministic knowledge of hydrological conditions.) This emphasizes the fundamental importance of hydrological forecasts, which can help reduce the losses and enhance the total effect of the control.

The utilization of forecasts in the control of reservoirs has received much attention in the water-engineering literature (e. g. [29, 62]). In our research we assessed the possibility of using short-term forecasts of affluences into a reservoir of the simplest statistical type. In forecasting average daily flows (for one day to one week) we made use of the linear regression relationships to the preceding days derived from a real (historical) flow series. Apart from the instantaneous application of the daily forecasts, we also considered the possibility of a several day's postponment of the decision concerning the change of

TABLE 32. A survey and brief characteristics of the models of short-term prognosis of the influx into the reservoir for the following day

Case	Characteristics of the model
P1	Control oriented towards constant minimum-plus runoff without any prognosis being applied
P2	Prognosis of the average influx into the reservoir for the following day of type $Q_{d+1}^{\mathrm{pr}} = b_1 Q_d + e_t$ (linear regression prognostic model of 1st order)
P3	Prognosis using the linear regression model of 1st order (comp. P2), but applied with a delay until 4th day from the issuing of the first adverse prognosis
P4	Prognosis of the average influx into the reservoir for the following day of type $Q_{d+1}^{\mathrm{pr}} = b_1 Q_d + b_2 Q_{d-1} + b_3 Q_{d-2} + e_t$ (linear regression model of 3rd order)
P5	Prognosis of the type $Q_{d+1}^{\mathrm{pr}} = Q_d$ based upon constant flow during dry periods poor in precipitation
P6	The so-called 100% successful prognosis of the type $Q_{d+1}^{\mathrm{pr}} = Q_{d+1}$ (a theoretical case of a prognosis in the form of the values of the average daily flows of the following days from the pre-modelled daily flow series)
P7	The so-called 100% successful advance prognosis for the whole dry period (deterministic case with complete a priori information on the course of the dry period)

operations, which was not to be made until a long-lasting adverse development of the hydrological situation had set in. The practical aim was to investigate the possibility of a less unstable regime of operations.

Table 32 presents a survey and the brief characteristics of the basic models of short-term prognosis of affluence for the immediately following day. For formal reasons, the basic variant of control oriented towards long-term constant delivery, with the prognosis with which the effectiveness of the other variants has been compared omitted, is denoted as P1.

The models of weekly forecasts were of an analogous character, again making use of linear regression for sectional investigation of the relationships between the neighbouring weekly affluences.

Beside short-term forecasts, we examined the possibility of taking advantage of a medium-term prognosis of water supply for a whole quarter of a year. In the case under examination, viz. the Orlík reservoir on the river Vltava, we were able to take advantage only of the prognoses for the third quarter of the year, since the correlation between the inflows in the other quarters of the year was rather weak.

Our research proceeded by the optimum control in real time being examined in a number of variants for various combinations of the decision and the forecasting models. In spite of the fact that the derivation of the optimum model with the help of long synthetic series of average daily flows may be numerically relatively demanding, experience has shown that the optimum can be approximated to very rapidly. The decision models themselves are simple enough and their practical application should pose no problems.

The decision models differed mutually by the measure to which they cut the deliveries in the dry years, this measure being virtually the expression of a stronger or weaker aversion to running the risks of economic losses. The decision model akin to model R according to Table 31, with short-term prognosis P5 (or even P6) made use of, proved to be the most suitable . In Table 33 the result of

TABLE 33. Effect of the optimum decision model on the result of the control
Short-term prognosis P5

Decision model	Total volume unsupplied		Total economic loss	
	($m^3 s^{-1}$)*)	(%)	(j)(**)	(%)
K	1 081	100	739	100
R	1 777	164	614	83
OPT	1 081	100	89	12

*) Sum of the deficits of guaranteed supply ΔO on the individual days of a 100-year synthetic series with respect to long-term value $O_p = 37 \text{ m}^3 \text{ s}^{-1}$.
**) In the units of loss function L1.

the solution is compared with both simple control oriented towards long-term constant delivery K without prognosis, and the deterministic approach to optimization, where according to Table 32 prognosis is also omitted and the reduction of delivery is optimized, with the course of the whole dry period a priori fully known, of course.

Table 33 shows that model R of control can help to cut the losses by 17 % as compared with variant K. The OPT variant must be regarded as a theoretical, practically unachievable, case. It is however useful in demonstrating the limits of control in the theoretical case where a 100% successful prognosis was available for the whole dry period.

If use were also made of a medium-term (quarterly) prognosis for a dry period beside prognosis P5, the losses would be cut by up to 22 % as compared with variant K. It follows that the effect of the quarterly prognosis can, in the given case, be assessed at an approximately 5 % cut in losses.

A detailed investigation of the set of the R decision models showed that a certain risk of economic losses due to the cuts in the deliveries of water is advisable and should be readily taken. Under the conditions of stochastic indeterminacy an aversion to that risk, manifesting itself in the controller's pessimism giving rise to substantial cuts in water deliveries, often leads to unnecessary losses in some of the failure-free periods, which can unfortunately usually not be offset by savings in the failure-prone dry periods.

Table 33 also shows that with adaptive control of the deliveries (and cuts in these deliveries) the economic losses can be reduced, but the volume of the

TABLE 34. Effect of the choice of the type of prognostic model P on the result of the control
Decision model R

Type of prognosis (accord. to Tab. 32)	Total volume unsupplied ($m^3 s^{-1}$)*)	(%)	Total economic loss (j)**)	(%)
P1	1 081	100	739	100
P2	1 657	153	746	101
P3	1 584	146	709	96
P4	1 227	135	769	105
P5	1 777	164	614	83
P6	1 775	164	614	83
P7	1 081	100	89	12

*) Sum of the deficits of guaranteed supply ΔO on the individual days of a 100-year synthetic series with respect to long-term value $O_p = 37\ m^3\ s^{-1}$.

**) In the units of loss function L1.

undelivered water may thus increase considerably. The other decision models investigated can substantially reduce the undelivered volume, but the economic losses will grow. The total effect of control is thus to a great extent dependent upon the criterion of optimality. There is no doubt that it is primarily the efficient prognostic models that can alleviate the effect of the controller's pessimism, particularly in the failure-free dry periods, when unnecessary cuts in the deliveries may often be imposed. For rational management it is thus of particular importance that efficient medium-term, and also long-term, forecasts should be attempted, which could help to upgrade decision-making in the whole dry period.

The effect of the choice of the P prognostic model on the result of the control is shown in Table 34. It is interesting to note that in the control of dry period deliveries the best variants are the models with the simplest type of forecast (P5, P6). Practical control of deliveries in real time would probably make use of the types of precipitation-and-runoff models, of the afflux line in the dry periods etc.

TABLE 35. Effect of the loss function on the result of the control

Type of loss function	Total economic loss (j) with control type		
	K	R*)	(%)
L1	739	614	83.2
L2	218	228	104.6
L3	65	90	138.5
L4	400	206	51.5
L5	170	67	39.4
L6	1 428	1 218	85.3
L7	714	609	85.3
L8	1 657	1 476	89.1
L9	2 237	1 650	73.8
L10	1 542	1 227	79.6

*) With prognostic model P5 applied.

Table 35 shows the dependence of the effect of adaptive control upon the type of the loss function, if prognosis P5 is employed. Adaptive control with model R is inferior to control K oriented towards long-term guaranteed delivery only with loss functions L2 and L3. This can be accounted for by the fact that these functions exhibit less pronounced skewness with which only relatively small reduction of the losses can be achieved while the reservoir is empty. (For with these loss functions it makes practically no difference whether the reduction of the economic losses is effected during a short and deep hydrological failure, or during a longer and shallower failure.) In the failure-free dry periods the losses will on the contrary rise as compared with control K, and during failures of the supply of water they can even exceed the economies achieved. Adaptive control need thus in no way always be effective.

Experience indicates that adaptive control can be effective wherever it will curb the high expected losses in critically dry periods. On the other hand, adaptive control loses its efficiency when the loss function is linear, or almost linear.*)

The relatively highest effects could be proved with loss functions L4 and L5. (According to Table 35, adaptive control, as compared with control K, can reduce losses by 39.4 to 51.5 %). This result can be accounted for by the fact that in these cases adequate measures can prevent economic losses caused by the smaller and quite frequent cuts in deliveries, which are the result of stochastic indeterminacy and the controller's pessimism in failure-free dry periods.

With the loss functions of type L6, L8, L9 and L10 the total absolute losses are considerably high throughout the whole flow series. They are influenced by a marked loss increment, or even by a jump in the growth of the losses, L9, L10, caused by a more radical cut in the deliveries of water. It is however worth mentioning that in these cases also, adaptive control proves more advantageous than the control oriented towards long-term constant delivery.

From this point of view it is also interesting that adaptive control using short-term prognoses is more advantageous in all the other variants examined, with the exception of the L2, L3 cases mentioned. Even though the total losses in dry periods can of course differ considerably according to the type of the loss function, the relative effects of the control (in terms of the percentage of the losses in constant delivery) are approximately the same. In this sense, adaptive control can be said to manifest certain robust properties owing to the shapes of the curves of the loss functions.

The results of the solution of adaptive control for the set of loss functions achieved also indicate that in concrete cases an approximate estimate can be made of the expected effect of the control with the help of the loss function computed.

Interesting results were also obtained from the investigation of a number of variants of the decision models of control with weekly statistical forecasts, which can be used separately or in connection with daily forecasts. If the daily forecast is checked against the total development of the hydrological situation for at least a period of a week, viz. with the help of a weekly forecast, unnecessary limitation of delivery due to pessimistic daily prognosis can often be prevented. The effect of control can thus be enhanced to some extent as compared with the case where only a daily forecast is taken into account. The effect is however greatly dependent upon the success of the forecast.

---

*) In water engineering, the concave loss function, with which the loss increment due to low cuts in deliveries is greater than the loss increment produced by a more radical cut, is also of no practical importance.

Figure 72 gives a schematic representation of ten basic cases of decision-making, arranged according to the success or failure of the two forecast. The diagram also visualizes the consequences of the decision made on the basis of erroneous forecasts.*)

Deeper investigation of the mechanism of these phenomena in the synthetic series of average daily flows will show that the merits and demerits of the two forecasts can manifest themselves variably, viz. in dependence upon the development of the hydrological conditions. The total effect of the two forecasts can thus be different in different sections of the flow series.

Research into control aided by simulation models also showed that with the step of a forecast extended, the success of that forecast grows less and the danger of uncertain or erroneous operations arises. For practical reasons, it is therefore desirable that perfection of the models with the longest possible head start of the forecast should be attempted, which could both reduce the frequency of operating interventions and upgrade decision-making throughout the whole dry period.

12.4.3 Properties of parameter estimates of adaptive control of seasonal reservoirs in real time

Research into the properties of the decision models for the control of reservoirs led us to the stochastic idea of the effects of control achieved as random variables, the behaviour of which in the long run depends upon the changeable hydrological conditions in different periods of time. The practical importance of this idea for decision-making, based on indeterminacy, can be demonstrated by the mutual relationship of the effects of control in various types of dry periods. If, in the failure-prone dry periods (ending in complete depletion of the reser-

---

*) In order to clarify the process of decision-making in various stituations let us consider two examples from Fig. 72.

Branch C: since the daily forecast is adverse, the development is controlled with the help of a weekly forecast, which is however favourable. No unnecessary limitation of the delivery is therefore undertaken. As both the forecasts have proved successful, the decision is rightly taken and the weekly forecast has had its full effect, viz. it has precluded unnecessary limitation of the delivery, which would have been decided upon if only the daily forecast had been taken into consideration.

Branch F: since the two forecasts are adverse, limitation of delivery has been decided upon. But as the weekly forecast is adverse, it has wrongly encouraged the imposition of the limiting regime decided upon on the basis of the daily forecast. The imposition of the limits has thus proved practically uselless and has led to unnecessary losses caused by inadequate control.

Conclusions:

1. If the weekly forecast is successful, its effect will manifest itself regardless of the fact whether the daily forecast is successful or not.

2. It is desirable and will prove effective that the most reliable forecast should be attempted, with the longest head start.

voir), the limited water delivery regime is switched to in time, this measure can reduce the economic losses due to reduced deliveries during the period of the failure itself. Unnecessary losses, on the other hand, are inflicted by the delivery being curbed in the less adverse dry periods that do not end in the reservoir being

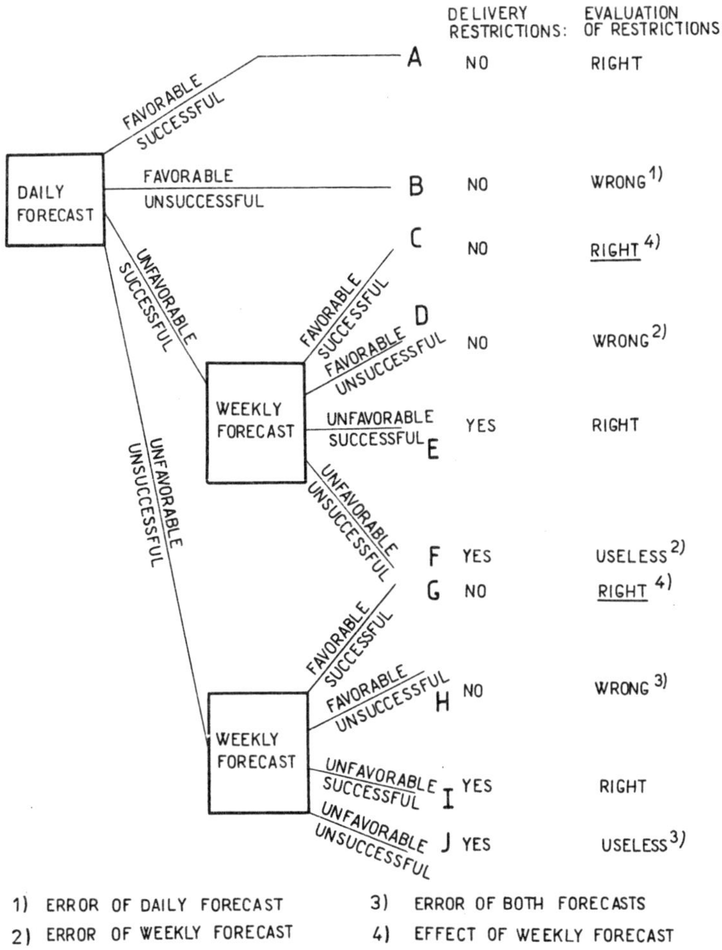

Fig. 72. The decision tree with the weekly forecast taken into account.

emptied completely. Unnecessary losses would not be inflicted if a long-term dependable forecast were available, justifying the reservoir controller's optimism, and the reservoir could thus be used without any limitation to guarantee long-term insured delivery.

The problem of taking decisions under indeterminacy led us to investigate the properties of the losses due to cuts in the delivery of water in various random realizations of the flow series and to estimate the effects of the control based upon these realizations. From the methodological point of view use is thus made of both the theory of statistical estimation and the theory of adaptive processes.

Research into these problems is extraordinarily difficult, for two reasons: the first serious difficulty is the investigation of the properties of the parameter estimates of the very short step flow series (e. g. average daily flow series), upon which the control in real time must inevitably be based. The second complex problem concerns the mutual relationships between the effects of adaptive control and the probability properties in various random realizations of the flow series.

The main methodological problem is the definition of the properties of control in dependence upon the changeable hydrological conditions, and also upon the other conditions of control. This dependence can be rewritten in the following form,

$$E = f(H,\, Q^{\mathrm{pr}}_{d+\tau},\, I,\, \mathrm{R},\, L,\, A_{\mathrm{z}}\, Q_{\mathrm{n}})\,, \tag{12.2}$$

where E is the effect of control expressed as the difference between the losses caused by limited deliveries of water under adaptive control and those consequent on the control oriented towards long-term guaranteed ensured delivery from the reservoir;

H denotes probability properties of the hydrological regime,

$Q^{\mathrm{pr}}_{d+\tau}$ the forecast of average daily afflux into the reservoir for the $d + \tau$ day,

I additional information on the state of the river basin,

R the decision model,

L the loss function,

A_{z} the storage volume of the reservoir,

Q_{n} the long-term guaranteed delivery from the reservoir.

Deriving function (12.2) in its explicit analytical form and finding its extreme is an extraordinarily difficult task in view of the fact that the effect of control, E, is conditional upon a large number of variables, some of which are of a stochastic character. The decision model, R, itself is also dependent upon several variables.

The derivation of function (12.2) and finding its optimum is however no less difficult even if the computation involves an existing reservoir, the parameters and the loss function of which are known. Equation (12.2) can in this case be rewritten as

$$E_{A_{\mathrm{z}},Q_{\mathrm{n}},L} = f(H,\, Q^{\mathrm{pr}}_{d+\tau},\, I,\, \mathrm{R})\,. \tag{12.3}$$

Relationship (12.3) was computed with some approximation with the help of the realizations of the flow series of various length generated from a 1000-year synthetic series of average daily flows. In the sets of these realizations (always with respect to their given length) the time-related variability of the parameters of control was then quite easy to monitor.

In each realization we monitored both the total economic losses caused by the limited delivery of water in the dry periods and the two components of these losses, viz. the losses in the month of the failure (with the reservoir completely depleted) and the losses caused by unnecessary limitation in the failure-free periods. The total losses as well as their components derived for the whole synthetic flow series, approximating to the population, were then compared with the losses in each realization. This methodological procedure, currently used in the theory of statistical estimation for the investigation of statistical sample characteristics, enabled us to ascertain the properties of the bias of the losses in the individual realizations.

Beside the properties of the losses in realizations, our main interest was in the probability properties of the flow series in the respective realizations and their possible correlations with the losses. Dealing with this proves to be rather difficult, because the losses have been derived from the average daily flows in each realization; for the set of these realizations it would also be necessary, for this purpose, to define their probability properties and project them in their aggregate on to the set of the "sample" losses, which are expressed by a single number for each realization. In order to assess this problem approximately at least, we expressed the probability properties of all the realizations using the statistical characteristics of the average monthly flows and monitored individualy the hydrologically adverse realizations that are linked with the highest control-inflicted losses.

We set out to process the parameters of control in thirty-year mutually independent realizations derived from a modelled 1 000-year series of average daily flows in the Orlík (Kamýk) profile of the river Vltava and obtained the first concrete results. In Table 36 the set of thirty-three realizations was divided into three groups according to whether they were linked with complete depletion of the reservoir, viz. a failure of water supply, or not. The group of the realizations linked with that failure was denoted as type A, the group not involving any failure but a limitation of the deliveries, as type B, and the group involving neither any failure nor a limitation of the deliveries of water as type C.

For the group of realizations of type A we then computed the total losses caused by the non-delivery of water in the course of thirty years, as well as the two components of these losses, viz. the losses linked with the failure itself and the losses caused by unnecessary limitation under the conditions of indeterminacy, thus unrelated to the failure. The group of realizations type B is characterized only by the losses caused by unnecessary limitation of water delivery, and group C by zero losses, because in these realizations the deliveries are ensured without any limitation (i. e. 100 percent).

From the point of view of the practical effect of control in real time, it is desirable not only to reduce losses in the failure-prone periods but also to

TABLE 36. Parameters of the control of the Orlík reservoir in 30-year realizations (loss function L1, decision model R2)

Realization No.	Realization period	Type of realization A B C	Loss in failure	Loss not in failure	Loss total	Indicator of losses excluding failures	Total number of days of limitation	Minimum fall in storage
1	11– 40	A	319	103	422	0.090	215	0
2	41– 70	A	181	0	181	0	20	0
3	71–100	C	0	0	0	0	0	1 814
4	101–130	C	0	0	0	0	0	2 078
5	131–160	C	0	0	0	0	0	2 908
6	161–190	C	0	0	0	0	0	1 800
7	191–220	B	0	1 386	1 386	1.212	488	0.8
8	221–250	C	0	0	0	0	0	2 033
9	251–280	C	0	0	0	0	0	2 415
10	281–310	C	0	0	0	0	0	1 091
11	311–340	C	0	0	0	0	0	2 165
12	341–370	C	0	0	0	0	0	1 915
13	371–400	B	0	386	386	0.337	253	55
14	401–430	B	0	117	117	0.114	138	557
15	431–460	C	0	0	0	0	0	1 373
16	461–490	C	0	0	0	0	0	2 805
17	491–520	C	0	0	0	0	0	1 785
18	521–550	A	1 144	457	1 601	0.399	788	0
19	551–580	B	0	137	137	0.120	161	126
20	581–610	C	0	0	0	0	0	2 081
21	611–640	B	0	120	120	0.105	124	797
22	641–670	C	0	0	0	0	0	2 063
23	671–700	A	966	350	1 316	0.306	596	0
24	701–730	C	0	0	0	0	0	1 533
25	731–760	C	0	0	0	0	0	1 716
26	761–790	B	0	344	344	0.301	274	439
27	791–820	B	0	40	40	0.035	41	241
28	821–850	C	0	0	0	0	0	1 807
29	851–880	B	0	92	92	0.080	116	511
30	881–910	A	643	86	729	0.075	306	0
31	911–940	C	0	0	0	0	0	1 783
32	941–970	C	0	0	0	0	0	788
33	971–1 000	A	1 113	33	1 146	0.029	219	0
Total for 990 years			4 366	3 651	8 017			
Average of all the realizations			132.3	110.6	242.9			
Average of 14 loss realizations			311.8	260.8	572.6			
Ratio of the highest loss in the realization to average loss of 14 realizations			3.67	5.31	2.80			

preclude unnecessary losses in the failure-free periods. We therefore attempted to find for these periods an indicator to characterize the measure of unnecessary losses related to the highest losses in the most adverse failure period in the set of all realizations, viz. in the whole 1 000-year series.

The literature dealing with the theory of games stresses the numerous difficulties in defining the optimum strategies under the given measure of indeterminacy of the conditions of control, as well as the fact that none of the principles, viz. definitions and indicators, is so convincing as to be accepted as a single and readily applicable basis of practical decision-making.

In order to express the magnitude of the necessary losses under the given measure of indeterminacy of the conditions of control, we introduced an indicator of the failure-unrelated losses (henceforth referred to as IFUL) as a dimensionless number defined by the ratio of the unnecessary losses caused by the limitation of the delivery of water in the failure-free periods to the highest losses in the failure situations correspondings to the optimum decision model.

According to this IFUL indicator, control in real time will be the more rational, the more that the indicator approximates to zero, i. e. the higher the preclusion of the reservoir controller's losses caused by the non-delivery of water. This can be achieved both by using the controller's experience and by the adoption of user-oriented measures, as well as the construction of more reliable predictive models.

The actual value of the indicator of failure-unrelated losses (IFUL) would be ascertainable by the unnecessary losses being evaluated after the end of the regime of delivery limitation (viz. after the end of the dry period), and the highest inevitable losses in failure situations are estimable in advance on the basis of a model solution. Since these highest losses in failure situations can, in a given profile, be regarded as constant, the values of the IFUL indicator fluctuate only in dependence on the unnecessary failure-unrelated losses.

The fluctuation of the IFUL indicator in the individual realizations (comp. Table 36) is quite considerable and it proves how sensitive control is to changeable hydrological conditions. For instance, the highest value of the indicator of failure-unrelated losses (1.212) is most interesting, for it shows that from the point of view of control failure-free realizations may occur, though more adverse than the realizations with the highest failure-related loss.[*)] This applies to the protracted, long-lasting dry periods, which can very often make the reservoir controller limit water delivery, unless of course the development of the hydrological conditions is well-known in advance.

---

*) From this point of view it can be claimed that the IFUL indicator expresses the hydrological conditions throughout the whole realization more directly and adequately than the classical long-term ensured delivery, which is determined on the basis of the most adverse section of the series in the given period of observation regardless of all the other periods.

In the other realizations the IFUL indicator did not exceed 40 percent, which is also rather high and points to the importance of rational management in dry periods.

This requirement is supported by the statistical characteristics of losses at the bottom of Table 36 numerically expressed for all the realizations. In the given case, the unnecessary losses amount to approximately 46 percent of the total losses. It is also worthy of note that the highest unnecessary loss in a realization can be much higher than the average loss. In the given case, the highest unnecessary loss is more than five times the average loss. This again confirms the need for systematic perfectioning of the predictive models, which can greathly contribute to more rational water management in dry periods.

The adaptive control of operations is extraordinarily efficient under the conditions where adequate measures can help to avoid economic losses caused by minor repeated cuts in water delivery. This is shown in Table 37, where the same analysis is applied to loss function L5. In spite of the fact that, according to that model variant of control, cuts in water delivery are also effected (again owing to the uncertainty of the development of the hydrological conditions), losses in failure-free periods have fallen off considerably. The highest IFUL value recorded in realization 7 reached 18 % only, the average of the unnecessary losses for all the realizations amounting to only 7 % of the total losses (compared with 46 % in the case of loss function L1). The highest loss in a realization can in this case however also be as much as five times the average loss, which is a proof of the rather variable character of the properties of the individual realizations and their effect on control.

Analogous properties of losses can also be proved for other types of decision models. Interesting results were obtained particularly from the examination of the relationships between the rate of the risk inherent in the process of decision-making and the losses caused by the control in dry periods. Unnecessary losses decrease with the risk of decision-making increasing, viz. in the cases where delivery starts being limited as late as in the end of the dry period marked with major depletion of reservoirs and the most unfavourable prognosis.

For the design of the operations it is however the total losses that are the most decisive, viz. the losses in both the failure-prone and the failure-free dry periods. In the case under examination, the Orlík reservoir on the river Vltava, the region of the optimum operations lies near decision model R, which can be characterized as mildly adverse to the risk of the occurrence of losses. It is however evident that this region of the optimum control has to be sought from case to case, because it depends upon a greater number of factors.

The distribution of the losses in the realizations is shown in Fig. 73, where the losses are ordered according to their magnitudes for the individual decision models and numbered according to the order of the realization. The growing index i of models R_i corresponds with the system of decision-making charac-

TABLE 37. Parameters of the control of the Orlík reservoir in 30-year realizations (loss function L5, decision model R2)

Realization		Type of realization			Loss			Indicator of losses excluding failures	Total number of days of limitation	Minimum fall in storage
No.	period	A	B	C	in failure	not in failure	total			
1	11– 40	A			34	0	34	0	28	0
2	41– 70	A			33	0	33	0	14	0
3	71–100			C	0	0	0	0	0	1 814
4	101–130			C	0	0	0	0	0	2 078
5	131–160			C	0	0	0	0	0	2 908
6	161–190			C	0	0	0	0	0	1 800
7	191–220		B		0	28	28	0.182	208	0.8
8	221–250			C	0	0	0	0	0	2 033
9	251–280			C	0	0	0	0	0	2 415
10	281–310			C	0	0	0	0	0	1 091
11	311–340			C	0	0	0	0	0	2 165
12	341–370			C	0	0	0	0	0	1 951
13	371–400		B		0	1	1	0.006	47	55
14	401–430			C	0	0	0	0	0	557
15	431–460			C	0	0	0	0	0	1 373
16	461–490			C	0	0	0	0	0	2 805
17	491–520			C	0	0	0	0	0	1 785
18	521–550	A			116	0	116	0	147	0
19	551–580			C	0	0	0	0	0	126
20	581–610			C	0	0	0	0	0	2 081
21	611–640			C	0	0	0	0	0	797
22	641–670			C	0	0	0	0	0	2 063
23	671–700	A			57	2	59	0.013	157	0
24	701–730			C	0	0	0	0	0	1 533
25	731–760			C	0	0	0	0	0	1 716
26	761–790		B		0	2	2	0.013	25	439
27	791–820			C	0	0	0	0	0	241
28	821–850			C	0	0	0	0	0	1 807
29	851–880			C	0	0	0	0	0	511
30	881–910	A			28	0	28	0	103	0
31	911–940			C	0	0	0	0	0	1 783
32	941–970			C	0	0	0	0	0	788
33	971–1 000	A			154	0	154	0	127	0
Total for 990 years					422	33	455			
Average of all the realizations					12.8	1.0	13.8			
Average of 9 loss realizations					46.9	3.7	50.6			
Ratio of the highest loss in the realization to average loss of 9 realizations					3.29	7.58	3.04			

terized by a rising risk of losses. The graphic representation highlights considerable fluctuation of the losses in the individual realizations with both the given decision model and the different decision models. It follows that a random

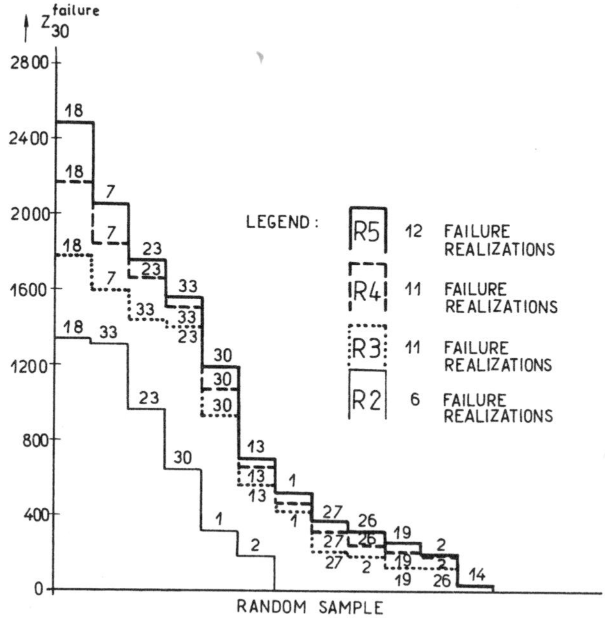

Fig. 73. Distribution of losses in the failure-prone periods of 30-year random realizations.

sample of a realization can yield random parameters of control considerably deviated from the long-term solution. This confirms the well-known requirement concerning the necessity of paying close attention to the representativeness of the initial hydrological data. However, the most important is the problem of the relationship between the variability of the parameters of control and the probability properties of the corresponding realizations. As already mentioned above, these relationships should in general be examined with the daily step, viz. with the parameters of control derived from the daily flow series and the statistical characteristics of these series. In order that this problem may be solved, an analysis would have to be made, among other things, of the variability of the statistical characteristics of the sets of the individual daily flows.

Instead, we made an attempt to clarify these relationships at least approximately by assessing the statistical characteristics of a set of thirty-three 30-year realizations of average monthly flows. In Fig. 74 are plotted the lines of transgression of the average long-term flows and the coefficients of variation and asymmetry in the winter months, when the demands on the control of the delivery from the Orlík reservoir on the river Vltava under examination are the highest. Table 38 lists the most significant statistical characterisistics of this set of realizations with the parameters of a 1 000-year synthetic series attached (always as the last lines), facilitating the assessment of systematic deviation.

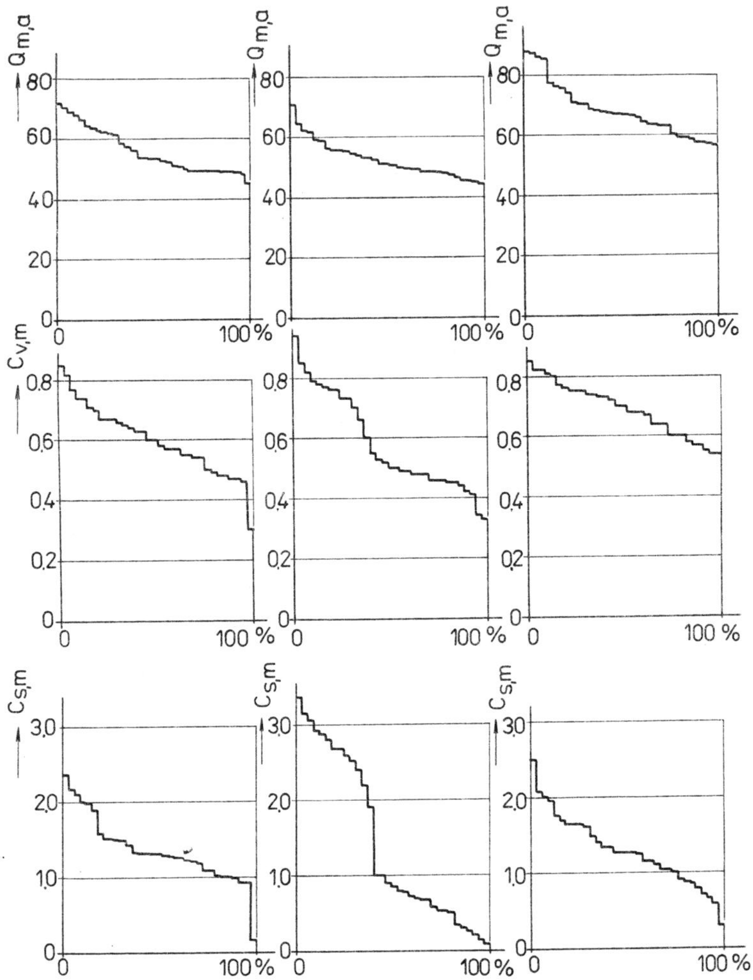

Fig. 74. Lines of transgression of average long-term flows, coefficients of variation and asymmetry of November, December and January flows in a set of 33 thirty-year random realizations in the Kamýk profile (Orlík) on the river Vltava (Czechoslovakia).

The course of the statistical characteristics in the lines of transgression, particularly the coefficients of variation and asymmetry, testifies to the fact that in that period of the year flows can fluctuate greatly and deviate considerably from their long-term values. It is thus evident that in the individual realizations

TABLE 38. Statistical characteristics of a set of 33 thirty-year realizations of average monthly flows in the Orlík profile of the river Vltava

Characteristics of the set of realizations	November flows	December flows	January flows
$Q_{ma,max}$	72.2	70.8	87.8
$Q_{ma,min}$	45.0	44.1	56.5
$E(Q_{ma})$	56.0	52.5	68.1
$C_v(Q_{ma})$	0.14	0.12	0.13
$C_s(Q_{ma})$	0.62	1.03	0.83
$Q_{ma,1000}$	56.0	52.5	68.1
$C_{v,max}$	0.85	0.94	0.85
$C_{v,min}$	0.30	0.33	0.54
$E(C_v)$	0.60	0.58	0.69
$C_v(C_v)$	0.19	0.28	0.13
$C_s(C_v)$	−0.08	0.49	−0.11
$C_{v,1000}$	0.64	0.63	0.70
$C_{s,max}$	2.37	3.36	2.52
$C_{s,min}$	0.16	0.08	0.30
$E(C_s)$	1.36	1.41	1.30
$C_v(C_s)$	0.32	0.78	0.36
$C_s(C_s)$	0.19	0.45	0.32
$C_{s,1000}$	1.85	2.96	1.47

$Q_{ma,max}$ ($Q_{ma,min}$) – maximum (minimum) average long-term flow in the set of realizations in the given month,

$E(Q_{ma})$ – mean of average long-term flows in the set of realizations in the given month,

$C_v(Q_{ma})$ – coefficient of variation of average long-term flows in the set of realizations in the given month,

$C_s(Q_{ma})$ – coefficient of asymmetry of average long-term flows in the set of realizations in the given month,

$Q_{ma,1000}$ – average 1000-year long-term flow in the given month,

$C_{v,max}(C_{v,min})$ – maximum (minimum) coefficient of variation of average monthly flows in the set of realizations in the given month,

$E(C_v)$ – mean of coefficients of variation of average monthly flows of the set of realizations in the given month,

$C_v(C_v)$ – coefficient of variation of coefficients of variation of average monthly flows of the set of realizations in the given month,

$C_s(C_v)$ – coefficient of asymmetry of coefficients of variation of average monthly flows of the set of realizations in the given month,

$C_{v,1000}$ – coefficient of variation of average monthly flows of a 1000-year series.

The statistical characteristics of coefficients of asymmetry have an analogous meaning.

the highly variable properties of the hydrological regime may also lead to different values of the parameters of reservoir-controlled runoff.

In spite of the fact that the analyses may give only an approximate idea of the mutual relationships betwen hydrological data and control, they clearly indicate the possible errors caused by random sampling of these data. It is obvious that designing reservoir-controlled runoff in real time and optimizing that runoff must be based upon reliable assessment of the representativeness of the initial data.

Our research dealt with the Orlík reservoir, for which seasonal runoff control is typical. The investigation of the relationships between the parameters of adaptive control and the parameters of the hydrological regime in short realizations of the flow series can be substantially more complex with long-term runoff control. This follows from the basic knowledge of the probabilistic solution of the function of storage reservoirs, the design parameters of which depend to a high degree upon the probability properties of the initial hydrological data. The computation of the function of storage reservoirs with the help of shorter realizations of the flow series can therefore be burdened with considerable random errors. Optimization of the control of these reservoirs in real time will require attention in further research.

The estimation of the parameters of adaptive control of reservoirs in real time gives rise to the problem of the sensitivy of control to the type of the flow series. Although a few partial contributions [48] dealing with these problems have appeared in the water-engineering journals, a systematic work answering the question of which of the two factors is more pronounced, or to what extent the control in real time is robust with respect to the two factors, is still lacking.

At present, optimization of the utilization of the water resources has become a very topical problem from both the short-term and operational point of view and the point of view of long-term non-stationary climatic variations that are expected to appear in the early decades of the next millennium. These problems have been considered by numerous international conferences (Villach 1985, Vancouver 1987), which pointed to the serious effects of such variations on the utilization of the water resources available. In this field, research will have to deal with both the assessment of the adaptivity of the existing resources to the expected future hydrological conditions and the estimation of the design parameters of new resources to be developed. Since in water engineering the period between the first scientific studies and the materialization of the respective measures adopted invariably equals 15 to 20 years, the urgency of research into these problems is steadily growing [49].

In this larger context, research into hydrological modelling will also need to be intensified. This applies particularly to the models with shorter time steps, which could help optimize utilization of the water resources with higher reliability.

12.5 Estimation of future climatic changes and their effect upon hydrologic regimes and water management in water resource systems

The estimation of global changes in the climate, and the effect of these changes upon hydrological regimes and the management of water resource systems, is an extraordinarily complex and hitherto unsolved problem of climatology, hydrology and water management. Since more pronounced climatic changes are to be expected as early as the beginning of next millenium, and since the preparation of the necessary water-engineering measures is invariably a long-term undertaking, this task has become highly topical. It is therefore most important that effective methods should be devised of upgrading the adaptability of the water resource systems and facing these changes.

The problems of future climatic changes have recently been dealt with by a number of international conferences and symposia, e. g. [122, 129, 130]. Some climatologists and hydrologists estimate that the content of carbon dioxide in the atmosphere may double by the beginning of next millennium if the contemporary trends continue. This may result in the intensification of the glasshouse effect and a rise in the average temperature by approximately 1.5 to 4.5 degrees centigrade. The effect of this rise upon the regional climates has however not been dependably analysed so far. It can however be safely assumed that the rise of temperatures will lead to higher variations of the climate and to the enhancement of climatic risks.

Another complex problem is posed by the effect of the expected climatic changes upon the hydrological regimes in the individual regions and river basins. The contemporary global climatologic models do not enable us to estimate the changes of the probability properties of flow series, which we need to know in order to deal with water resources and their systems. It is quite obvious that the climatological and hydrological data acquired by measurment in the past can in no way serve as a satisfactory basis for the estimation of the statistical characteristics of the flow series in the future. The assumption concerning the stationarity of these series will therefore have to be revised.

The changes in the climate and the impact of these changes upon the hydrological regimes will of course not manifest themselves as abrupt "jumps", but gradually and incrementally, dependent upon man's activity. Some hydrologists point out that quantification of such gradual changes from their contemporary onset till their "final" state at the beginning of next millennium is an extraordinarily complex problem. And in this context it will also prove necessary that the statistical significance of the non-stationary changes should be considered and compared with the changes admissible in the case of variation of stationary quantities [49, 123].

It thus follows that in order that water resources may be most rationally managed in the nearest period, the statistical characteristics of the hydrological series will have to be estimated in view of the changes in climate expected. Since it can be assumed that the contradiction between the hitherto obtaining assumption of stationarity of the hydrological series and reality will soon be deepening, the methodology of hydrological and water-engineering computations will have to be enriched and generalized to be also applicable to the non-stationary hydrological processes.

Since these problems are of a highly complex character, close cooperation between climatologists and hydrologists worldwide will be most expedient.

At the Faculty of Civil Engineering of the Czech Technical University in Prague research into these problems has been carried on along two lines. The first is the research concerning the mathematical model of irrigation requirements serving primarily for planning and designing the irrigation schemes in water resource systems. The model of irrigation requirements is based upon the monthly balance of the soil's moisture, and the value of potential evapotranspiration is calculated with the help of Penman's formula. The model is calibrated using a shorter period for which the meteorological data and irrigation requirements are available. It can then be applied to mathematization of the whole systems of water management. The model has been further elaborated, and its stochastic variant has been derived serving to generate random series of irrigation requirements concerning water. These irrigation requirements can be used as input values in stochastic simulation models [124, 125, 126]. The possibility is also studied of applying this approach to the estimation of irrigation requirements with respect to climatic changes.

The second line of the research carried out by the researchers of the Faculty of Civil Engineering consists in the investigation of the long-term changes and the variation of the probability properties of flow series. As it turned out, the hitherto generally accepted assumption of stationarity did not prove to be fulfilled in all the flow series examined. In some cases, the variability of sample characteristics (e. g. coefficients of variation or asymmetry) in time was proved. Most interesting are particularly the cases of periodic tendencies in the fluctuation of these characteristics, which form the basis for their extrapolation into the future.

At the next stage the researchers at the Faculty of Engineering of the Czech Technical University in Prague will attempt to estimate the contemporary occurrence of seasonal fluctuation with long-term changes. It is the aim of the research to derive a variant simulation model of flow series enabling at least an approximate estimate of the effect of climatic changes upon the management of water resources.

13
Prospects of the development of estimation theory

As far as the tendencies and prospects of the development of the theory of estimation and its applications are concerned, it is to be expected that in the near future their importance will grow, not only because new knowledge in the field of the theory of probability is rapidly appearing, but also because the tasks of hydrology and water engineering are getting more complex and exacting. In this respect the theory of estimation touches on a number of important problems, the solution of which makes the application of the probability approach indispensable.

For several decades the large palette of the methods of hydrological and water-engineering computations has been based upon a simplified assumption of the representativeness of the given series of hydrological quantities. The contemporary theory of estimation makes it possible for the relationships between samples and the universe hitherto completely unknown to be expressed and, if need be, for the parameters of the universe to be estimate.

However marked this progress may be, an important fact should not be overlooked, which is that the estimation of parameters is often based upon an assumption of the type of probability distribution of the population, which is actually completely unknown. And it is this assumption that the former indeterminacy of the assessment of the representativeness of the real sample and its characteristics has at present virtually shifted into.

The development of theoretical inquiry has therefore in no way been concluded as yet, as is shown by the work of the American statistician P. J. Huber on robust estimation in statistics [37], the Russian translation of which came out only after we had completed the Czech manuscript of the present work (Sept. 1984). On the basis of the literature available it can be claimed that P. J. Huber's is the first systematic treatise on the theory of robust estimation, which is intensively being developed as a trend in contemporary mathematical statistics. Hydrology and water engineering is thus facing a new and exacting task. The methods of monitoring the stability of statistical procedures, as well as the algorithms of the computations of robust estimates, will have to be scrutinized as far as the various types of hydrological series are concerned, and the conclusions drawn applied to the solution of the water-engineering problems. It is evident that without this work, the essence of the problems of parameter estimation can hardly be approached.

The various theoretical problems have been dealt with in the respective chapters of the present book. In their attemps at solving them the statisticians will now have to link much more consistently the statistical with the genetic methods, particularly in estimating the parameters of a system of dependent stations, the knowledge of which is indispensable as far as their mathematical modelling is concerned.

Perfecting the statistical and the genetic methods, as well as their interaction, will also have to be attempted in the solution of the complex of the important problems linked with the effect of the anthropogenic factors on the runoff regime and its expected development. These are pressing problems in view of the fact that the measure of man's activities influencing runoff is on the increase and the number of the flow series, or sections of flow series, unaffected by man's activity is rapidly decreasing. These changed conditions will have to be taken into account whenever the expected probability properties of the runoff regimes are estimated and their effects upon rational water management dealt with.

In order that these tasks may be successfuly tackled, it is above all essential that both the quantitative and the qualitative influence of the individual factors should be monitored, measured and evaluated, so that their effect may become ascertainable. It will be advisable to continue the research already started and elaborate the methods of solving these problems for the individual river-basins, so that those effects may be duly considered in the homogenization of the flow series ant their utilization in current water-engineering practice.

The importance of the problem of the effect of man's activity on the runoff regimes of rivers has recently been emphasized by international cooperation in this field of research, which started taking shape towards the end of the International Hydrological Decade. And within the framework of the International Hydrological Project, attention was also given to research into the anthropogenic changes in the water resources of the Earth. Since 1977 regional cooperation between the European ex-socialist countries in this research has greatly intensified; this cooperation has particularly concentrated upon the assessment of the effect of urbanization, town planning, farming and the construction of water reservoirs upon the hydrological regime and the quality of water [63].

In water management, the application of mathematical modelling has long proved to be a constant problem. For the flow series, mathematical models have satisfactorily been elaborated down to the monthly interval. As far as the shorter intervals are concerned, research will have to be continued. And for these series with shorter time intervals the possibility will also have to be examined of their parameters being estimated and brought on to the input of the respective models.

Some of the problems of mathematical modelling of the flow series still await full clarification as far as their applicability to the systems of water management is concerned, where synthetic series are to be modelled for a whole system of stations and with the mutual relationships of correlation taken duly into account.

The theory of systems of water management belongs to the fields of research where the theory of estimation is applied to both the processing of hydrological data and mathematical modelling of the water management systems itself, which is a particularly useful tool for designing these systems and utilizing them optimally. The development of the theory of systems of water management is marked with various applications of systems approaches of optimization, the most topical of which is the approach of the statistical theory of decision-making, particularly under the conditions of risk and indeterminacy (e. g. incomplete information), which must of course be fully considered if optimum control of the systems in real time is to be achieved.

All of these problems are both exacting and comprehensive, they can therefore not be completely resolved without systematic research. It can be assumed that research into these problems will also create favourable conditions for the substantial extension of the field of application of the theory of estimation.

Bibliography

[1] ALEKSEEV, G. A.: Graphical and Analytical Methods of Determination of Sample and Population Parametrs of Distribution Functions. Gidrometeoizdat, Leningrad 1960, No. 73, p. 90–140. (АЛЕКСЕЕВ, Г. А.: Графоаналитические способы определения и приведения к длительному периоду наблюдений параметров кривых распределения, in Russian).

[2] ANDĚL, J.: Statistical Analysis of Time Series. SNTL, Prague 1976, 272 p. (Statistická analýza časových řad, in Czech).

[3] ANDĚL, J.: Mathematical Statistics. SNTL/ALFA, Prague 1978, 352 p. (Matematická statistika, in Czech).

[4] ANDĚL, J., BALEK, J.: Modelling of Hydrological Series. Institute of Hydrodynamics of the Czechoslovak Academy of Sciences, Prague 1969, Report No. 225/D/69, 36 p. plus append. (Modelování hydrologických řad, in Czech).

[5] ANDĚL, J., BALEK, J.: Mathematical and Statistical Method of Analysis of the Generation of Hydrological Series. Hydrological Journal, **18**, 1970, No. 1, p. 3–28 (Matematicko-statistická metoda analýzy tvorby hydrologických řad, in Czech).

[6] ANDERSON, O. D.: Time Series Analysis and Forecasting – The Box–Jenkins Approach. Butterworth, London 1976.

[7] BALEK, J.: Linear Extrapolation of Average Annual Flows of Selected Rivers of Four Continents. Hydrological Journal, **16**, 1968, No. 3, p. 402–428 (Lineární extrapolace průměrných ročních průtoků vybraných řek čtyř kontinentů, in Czech).

[8] BARTLETT, M. S.: On the Theoretical Specification of Sampling Properties of Autocorrelated Time Series. J. Royal Statist. Soc., B 8, 1946, p. 27–41.

[9] BLOKHINOV, E. G.: New Methods of Estimation of Parameters of the Fluctuations of Annual Flows on the Basis of Long-term Observation. Gidrometeoizdat, Leningrad 1968, No. 143, p. 134–185 (БЛОХИНОВ, Е. Г.: Новые приемы оценки параметров случайных колебаний речного стока по данным многолетних наблюдений, in Russian).

[10] BLOKHINOV, E. G., SOTNIKOVA, L. F.: On the Estimation of Parameters of Probability Distribution of the Annual Flows of the Rivers of the USSR. Gidrometeoizdat, Leningrad 1970, No. 180, p. 85–123 (БЛОХИНОВ, Е. Г., СОТНИКОВА, Л. Ф.: Об оценке параметров распределения вероятностей годового стока рек СССР, in Russian).

[11] BOBÉE, B.: The Log Pearson Type 3 Distribution and Its Application in Hydrology. Wat. Resour. Res., **11**, 1975, No. 5, p. 681–689.

[12] BOBÉE, B., ROBITAILLE, R.: Correction of Bias in the Estimation of the Coefficient of Skewness. Wat. Resour. Res., **11**, 1975, No. 6, p. 851–854.

[13] BOBÉE, B., ROBITAILLE, R.: The Use of the Pearson Type 3 and Log Pearson Type 3 Distributions Revisited. Wat. Resour. Res., **13**, 1977, No. 2, p. 427–443.

[14] BOX, G. E. P., JENKINS, G. M.: Time Series Analysis, Forecasting and Control. Holden Day, San Francisco 1970.

[15] BRATRÁNEK, A.:Long-term Forecasts of Flows on Rivers and Their Importance for the Economical Operation of Water Engineering Works. Práce a studie VÚV, Prague 1962, No. 109, 72 p. (Dlouhodobé předpovědi průtoků na tocích a jejich význam pro hospodárný provoz vodních děl, in Czech).

[16] BRATRÁNEK, A.: Solar Activity and Its Effect on the Fluctuation of Hydrological Phenomena. Práce a studie VÚV, Prague 1965, No. 117, 84 p. (Sluneční aktivita a její vliv na kolísání hydrologických jevů, in Czech).

[17] BRATRÁNEK, A.: Variability of Flows and Coefficient of Variation in 100-year Flow Series. Hydrological Journal, **14**, 1966, No. 1, p. 3–19 (Proměnlivost průtoků a součinitele variace ve stoletých průtokových řadách, in Czech).

[18] BROŽA, V.: Symposium on the Methods of Runoff Control by Reservoirs. Water Management, 1975, No. 1, p. 13–14 (Sympozium o metodách řízení odtoku nádržemi, in Czech).

[19] BROŽA, V., NACHÁZEL, K., VITHA, O.: Statistical Research Into the Regularities of the Flood Regime in Smaller Streams. Hydrological Journal, **26**, 1978, No. 1, p. 3–33 (Statistický výzkum zákonitostí povodňového režimu malých vodních toků, in Czech).

[20] BUBENÍČKOVÁ, L., KAŠPÁREK L.: Comparison of the Representativeness of Flow Series 1931–70 and 1931–60. HMÚ, Prague 1976, 171 p. and encl. (Porovnání reprezentativnosti průtokových řad 1931–70 a 1931–60, in Czech).

[21] BUCHTELE, J.: Analysis of Hydrological Series for Forecasting Seasonal Inflows Into Reservoirs. HMÚ, Prague 1975 (Analýza hydrologických řad pro předpovědi sezónních přítoků do nádrží, in Czech).

[22] BURGES, J. S., HOSHI, K.: Approximation of Normal Distribution by a Three - Parameter Log Normal Distribution. Wat. Resour. Res., **14**, 1978, No. 4, 620–622.

[23] BUŘITA, J. et al.: Rational Control of Water Resource Systems. Report on Research in 1983 National Research Project II–5–6/6, Prague 1983, 37 p. (Racionální řízení vodohospodářských soustav, in Czech).

[24] CIDLINSKÝ, J.: Rationalization of the Utilization of the Storage Capacity of a Reservoir. Final Study in the Postgraduate Course at the Faculty of Civil Engineering of the Czech Technical University, Prague 1982, 29 p. (Racionalizace využívání zásobního prostoru nádrže, in Czech).

[25] CIPRA, T.: Analysis of Time Series With Applications to Economy. SNTL–ALFA, Prague 1986, 248 p. (Analýza časových řad s aplikacemi v ekonomii, in Czech).

[26] CLARKE, A. B., DISNEY, R. L.: Probability and Random Processes for Engineers and Scientists. John Wiley and Sons, New York – London – Sydney – Toronto 1970, 346 p.

[27] CONDIE, R.: The Log Pearson Type 3 Distribution. The T-Year Event and Its Asymptotic Standard Error by Maximum Likelihood Theory. Wat. Resour. Res., **13**, 1977, No. 6, p. 987–991.

[28] Czechoslovak National Standard No. 73 6805 – Hydrological Data of Surface Waters. Prague 1975 (Hydrologické údaje povrchových vod, in Czech).

[29] DATTA, B., HOUCK, M. H.: A Stochastic Optimization Model for Real-Time Operations of Reservoirs Using Uncertain Forecasts. Wat. Resour. Res., **20**, 1984, No. 8, p. 1039–1046.

[30] DUB, O., NĚMEC, J. et. al.: Hydrology. SNTL, Prague 1969, 380 p. (Hydrologie, in Czech).

[31] FEDOROV, L. T.: On the Estimation of the Fluctuation of Annual Flows of Rivers in the Territory of Kazakhstan Using the Method of Maximum Likelihood. Gidroprojekt, Collection of Papers 4, Moscow 1960, p. 114–117 (Федоров, Л. Т.: Об оценке изменчивости годового стока рек на территории Казахстана методом наибольшего правдоподобия, in Russian).

[32] FIERING, M. B.: Queuing Theory and Simulation in Reservoir Design. J. Hydraulics Div. ASCE, HY 6, November 1961.

[33] GOULD, B. W.: Statistical Methods for Estimating the Design Capacity of Dams. Journ. Inst. Engs. Aust., **33**, 1961.

[34] GRINEVICH, G. A., PETELINA, N. A., GRINEVICH, A. G.: Structural Modelling of Hydrographs. Nauka, 1972, 181 p. (ГРИНЕВИЧ, Г. А., ПЕТЕЛИНА, Н. А., ГРИНЕВИЧ, А. Г.: Композиционное моделирование гидрографов, in Russian).

[35] HÁTLE, J., LIKEŠ, J.: Fundamentals of Probability Calculus and Mathematical Statistics. SNTL – ALFA, Prague – Bratislava 1972, 464 p. (Základy počtu pravděpodobnosti a matematické statistiky, in Czech).

[36] HAZEN, A.: Storage to be Provided in Impounding Reservoirs for Municipal Water Supply. Trans. of ASCE, Vol. 77, 1914, p. 1539–1669.

[37] HUBER P. J.: Robust Statistics. John Wiley and Sons, New York 1981.

[38] CHARBENEAU, J. R.: Comparison of the Two- and Three-Parameter Log Normal Distribution. Used in Streamflow Synthesis. Wat. Resour. Res., **14**, 1978, No. 1, p. 149–150.

[39] JAGLOM, A. M.: General Theory of Stationary Random Functions. Soviet Science, V, 1955, p. 108–139 (Obecná teorie stacionárních náhodných funkcí, in Czech).

[40] JENKINS, G. M., WATTS D. G.: Spectral Analysis and Its Applications. Holden Day, San Francisco, Cambridge, London, Amsterdam 1969.

[41] KARTVELISHVILI, N. A.: Stochastic Hydrology. Gidrometeoizdat, Leningrad 1975, 164 p. (КАРТВЕЛИШВИЛИ, Н. А.: Стохастическая гидрология, in Russian).

[42] KAŠPÁREK, L.: On Floods on the River Litavka in 1872 and 1981 and Their Importance for the Estimation of the n-Year Flows. ČHMÚ, Prague 1984, No. 7, 56 p. (O povodních z let 1872 a 1981 na Litavce a jejich význam pro odhad n-letých průtoků, in Czech).

[43] KAŠPÁREK, L.: Analysis of the Probability Properties of Hydrological Quantities and Their Mutual Relationships. Dissertation, Prague 1986, 158 p. plus append. (Analýza pravděpodobnostních vlastností hydrologických veličin a jejich vzájemných vztahů, in Czech).

[44] KAŠPÁREK, L. et. al.: Research into the Methods of Automated Data Processing of Hydrological Design Quantities. Partial Report on Project II–7–2/8 of the National Programme of Basic Research, HMÚ, Prague 1978, 19 p. (Výzkum metod automatizovaného zpracování návrhových hydrologických veličin, in Czech).

[45] KAŠPÁREK, L. et. al.: Selection of Methods of Automated Data Processing of the N-Year Flows. Final Report on National Research Project II–7–2/8, ČHMÚ, Prague 1980, 116 p. (Výběr metod automatizovaného zpracování N-letých průtoků, in Czech).

[46] KAŠPÁREK, L. et. al.: Methodology of Processing Series of Culminating Flows in ČHMÚ. Description of Computing Program. Final Report on Enterprise Research Project No. 143, ČHMÚ, Prague 1982, 24 p. plus append. (Metodika zpracování řad kulminačních průtoků v ČHMÚ. Popis výpočetního programu, in Czech).

[47] KLEČKA, V.: Rationalization of Water Management in Reservoirs. Final Study in the Postgraduate Course at the Faculty of Civil Engineering of the Czech Technical University, Prague 1983, 31 p. (Racionalizace hospodaření s vodou v nádržích, in Czech).

[48] KLEMEŠ, V.: Value of Information in Reservoir Optimization. Wat. Resour. Res., **13**, 1977, No. 5, p. 837–850.

[49] KLEMEŠ, V.: Sensitivity of Water Resource Systems to Climate Variations. World Climate Programme 98, World Meteorological Organization, Geneva May 1985.

[50] KLIBASHEV, K. P., GOROSHKOV, I. F.: Hydrological Computations. Gidrometeoizdat, Leningrad 1970, 460 p. (Клибашев, К. П. Горошков, И. Ф.: Гидрологические расчеты, in Russian).

[51] COLLECTIVE OF AUTHORS: Hydrological Regimes of the Czechoslovak Socialist Republic, Part III. HMÚ, Prague 1970, 305 p. plus append. (Hydrologické poměry ČSSR, in Czech).

[52] COLLECTIVE OF AUTHORS: Applied Mathematics, Part I and Part II. SNTL, Prague 1978, 2386 p. (Aplikovaná matematika, I. a II. díl, in Czech).

[53] KORN, G. A.: Random Process Simulation and Measurement. McGraw – Hill Book Co., New York – Toronto – London – Sydney 1966, 234 p.

[54] KOS, Z.: Determination of the Coefficient of Variation Using the Method of Maximum Likelihood. Water Management, 1967, No. 6, p. 241–243 (Určování koeficientu variace metodou maximální věrohodnosti, in Czech).

[55] KOS, Z.: The Linear Regression Model and Its Applications in Hydrology. Práce a studie, Vodní toky, Prague 1969, No. 6, 122 p. plus append. (Lineární regresní model a jeho aplikace v hydrologii, in Czech).

[56] KOS, Z.: Probability Models of Water Resource Systems. Práce a studie, VÚV, Prague 1978, No. 150/A, 189 p. (Pravděpodobnostní modely vodohospodářských soustav, in Czech).

[57] KOS, Z., ZEMAN, V.: Water Resource Systems in the Guiding Water Management Plan. SZN, Prague 1976, 271 p. (Vodohospodářské soustavy ve Směrném vodohospodářském plánu, in Czech).

[58] KRITSKIĬ, S. N., MENKEĽ, M. F.: Long-term Control of Flow. Gidrot. Stroit., 1935, No. 11 (КРИЦКИЙ, С. Н., МЕНКЕЛЬ, М. Ф.: Многолетнее регулирование стока, in Russian).

[59] KRITSKIĬ, S. N., MENKEĽ, M. F.: On the Application of the Method of Maximum Likelihood to Sampling Estimation of Statistical Parameters of River Flows. Izvestija AN USSR, Department of Technology, 1949, No. 4 (КРИЦКИЙ, С. Н., МЕНКЕЛЬ, М. Ф.: О применении метода наибольшего правдоподобия к выборочной оценке статистических параметров речного стока, in Russian).

[60] KRITSKIĬ, S. N., MENKEĽ, M. F.: Computation of Long-term Runoff Control with Respect to the Relationship of Correlation between Runoff in the Dry Years. AN USSR, 1959, No 8 (КРИЦКИЙ, С. Н., МЕНКЕЛЬ, М. Ф.: Расчет многолетнего регулирования стока с учетом коррелативной связи между стоком смежных лет. Проблемы регулирования речного стока, in Russian).

[61] KRITSKIĬ, S. N., MENKEĽ, M. F.: Hydrological Fundamentals of River Flow Control. Izdat. Nauka, Moscow 1981, 256 p. (КРИЦКИЙ, С. Н., МЕНКЕЛЬ, М. Ф.: Гидрологические основы управления речным стоком, in Russian).

[62] KRZYSZTOFOWICZ, R., WATADA, L. M.: Stochastic Model of Seasonal Runoff Forecasts. Wat. Resour. Res., **22**, 1986, No. 3, p. 296–302.

[63] KŘÍŽ, V.: Hydrological Regimes of Rivers and Their Changes Caused by Anthropogenic Effects. Doctoral Dissertation. ČHMÚ, Ostrava 1982 (Vodní režim řek a jeho změny působené antropogenními vlivy, in Czech).

[64] KUBÁČEK, L.: Fundation of Estimation Theory. Elsevier, Amsterdam and Veda, Bratislava 1987, VI + 328 p.

[65] LIKEŠ, J., MACHEK, J.: Mathematical Statistics. SNTL, Prague 1983, 180 p. (Matematická statistika, in Czech).

[66] LLOYD, E. H.: A Probability Theory of Reservoirs with Serially Correlated Inputs. Journ. of Hydrology, 1963, No. 1.

[67] MAŇAS, M.: Theory of Games and Optimum Decision-Making. SNTL, Prague 1974, 256 p. (Teorie her a optimální rozhodování, in Czech).

[68] MATALAS, N. C., WALLIS, J. R.: Eureka! It Fits a Pearson Type 3 Distribution. Wat. Resour. Res., **9**, 1973, No. 2, p. 281–289.

[69] MORAN, P. A. P.: A Probability Theory of Dams and Storage Systems. Aust. Journ. Appl. Sci., 1954, No. 5.

[70] MORAN, P. A. P.: A Probability Theory of Dams and Storage Systems: Modifications of the Release Rule. Aust. Journ. Appl. Sci., 1955, No. 6.

[71] NACHÁZEL K.: Relationships of Correlation in the Control of Runoff with the Help of Water Reservoirs. Doctoral Dissertation, Prague 1965, 141 p. (Korelační vztahy při regulování odtoku vodními nádržemi. In: Sborník Přehradní dny, 1965, in Czech).

[72] NACHÁZEL K.: Solution of Water Engineering Problems in a Set of Realizations of Random Flow Series. Water Management, 1976, No. 9, p. 229–232 (Řešení vodohospodářských úloh v souboru realizací náhodných průtokových řad, in Czech).

[73] NACHÁZEL, K.: Effects of Non-Stationary Hydrological Regimes in the Computation of Reservoir Design. Hydrological Journal, **24**, 1976, No. 1, p. 1–21 (Důsledky nestacionárních hydrologických režimů na řešení nádrží, in Czech).

[74] NACHÁZEL, K.: Stochastic Processes and Methods in Hydrology. Textbook for the UNESCO International Postgraduate Course in Hydrology. Prague 1978, 134 p. plus append.

[75] NACHÁZEL, K.: Statistical Research of Regularities of Sample Characteristics of Hydrological Series. Hydrological Journal, **28**, 1980, No. 3, p. 257–285 (Statistický výzkum zákonitostí výběrových charakteristik hydrologických řad, in Czech).

[76] NACHÁZEL, K.: Random, Probable and Systematic Errors of Estimation of Parameters of Hydrological Series. Hydrological Journal, **29**, 1981, No. 1, p. 9–19 (Náhodné, pravděpodobné a systematické chyby odhadu parametrů hydrologických řad, in Czech).

[77] NACHÁZEL K.: Problems of Estimation of Statistical Parameters of Hydrological Series by the Method of Maximum Likelihood and Regularities of Their Sample Characteristics, Hydrological Journal, **29**, 1981, No. 2, p. 113–136 (Problematika odhadu statistických parametrů hydrologických řad metodou maximální věrohodnosti a zákonitostí jejich výběrových charakteristik, in Czech).

[78] NACHÁZEL, K.: Problems of Bias of Statistical Characteristics of Average Monthly Flow Series and Their Mathematical Models. Hydrological Journal, **32**, 1984, No. 1, p. 3–31 (Problematika vychýlení statistických charakteristik řad průměrných měsíčních průtoků a jejich matematických modelů, in Czech).

[79] NACHÁZEL, K. et al.: Stochastic Models of Runoff Fluctuation During One Year and Their Effect on Rational Utilization of Water Resources. Final Report of Partial Project II–7–2/15 of the National Plan of Basic Research. Department of Hydrotechnology of the Czech Technical University, Prague 1975, 87 p. (Stochastické modely kolísání odtoku během roku a jejich vliv na racionální využití vodních zdrojů, in Czech).

[80] NACHÁZEL, K. et al.: Problems of Estimation of Statistical Parameters of Hydrological Series by the Method of Maximum Likelihood and the Regularities of Their Sample Characteristics. Final Report. Department of Hydrotechnology of the Czech Technical University, Prague 1980, 40 p. and append. (Problematika odhadu statistických parametrů hydrologických řad metodou maximální věrohodnosti a zákonitostí jejich výběrových charakteristik, in Czech).

[81] NACHÁZEL, K. et al.: Research into the Effect of Extreme Values on the Magnitude of Bias of Sample Characteristics of Hydrological Series. Final Report. Department of Hydrotechnology of the Czech Technical University, Prague 1981, 19 p. and append. (Výzkum vlivu extrémních hodnot na velikost vychýlení výběrových charakteristik hydrologických řad, in Czech).

[82] NACHÁZEL, K. et al.: Mathematical Modelling of Average Monthly Flow Series with Respect to Bias of Statistical Characteristics. Final Report. Department of Hydrotechnology of the Czech Technical University, Prague 1984, 21 p. plus append. (Matematické modelování řad průměrných měsíčních průtoků se zřetelem k vychýlení statistických charakteristik, in Czech).

[83] NACHÁZEL, K., BUREŠ, P.: Computation of the Design of Carryover Reservoirs with the Help of Monte-Carlo Methods. Hydrological Journal, **21**, 1973, No. 1, p. 3–32 (Řešení víceletých nádrží metodami Monte-Carlo, in Czech).

[84] NACHÁZEL K., PATERA, A.: Statistical and Genetic Properties of Monthly Flow Series. Hydrological Journal, **20**, 1972, No. 6, p. 605–640 (Statistické a genetické vlastnosti měsíčních průtokových řad, in Czech).

[85] NACHÁZEL, K., PATERA, A.: Correlative and Spectral Properties of Hydrological Series. Hydrological Journal, **23**, 1975, No. 1, p. 3–35 (Korelační a spektrální vlastnosti hydrologických řad, in Czech).

[86] NACHÁZEL, K., PATERA, A.: Non-Stationarity of Hydrological Regimes. Hydrological Journal, **23**, 1975, No, 6, p. 527–561 (Nestacionarita hydrologických režimů, in Czech).

[87] NACHÁZEL, K., PATERA, A.: Effect of the Bias of Statistical Characteristics of Flow Series on the Computation of the Storage Function of Reservoirs. 1st Part: The Long-term Stationarity Function of Reservoirs. Hydrological Journal, **32**, 1984, No. 2, p. 113–138 (Vliv vychýlení statistických charakteristik průtokových řad na řešení zásobní funkce nádrží. 1. část: Dlouhodobá stacionární funkce nádrží, in Czech).

[88] NACHÁZEL, K., PATERA, A.: Effect of the Bias of Statistical Characteristics of Flow Series on the Computation of the Storage Function of Reservoirs. 2nd Part: Designing Reservoirs with the Help of Short Realizations of Flow Series. Hydrological Journal, **32**, 1984, No. 3, p. 243–267 (Vliv vychýlení statistických charakteristik průtokových řad na řešení zásobní funkce nádrží. 2. část: Řešení nádrží v krátkých realizacích průtokových řad, in Czech).

[89] OZAKI, T.: On the Order Determination of ARIMA Models. Appl. Statistics, **26**, 1977, p. 290–301.

[90] PATERA, A.: Computation of Adaptivity of Reservoirs with the Help of Short Realizations of Hydrological Processes. Hydrological Journal, **26**, 1978, No. 3, p. 228–244 (Řešení adaptivity nádrží v krátkých realizacích hydrologických procesů, in Czech).

[91] PROCHÁZKA, M.: Comparison of Various Methods of Determination of the Order of Autoregression Model in the Modelling of Average Monthly Flows. Hydrological Journal, **32**, 1984, No. 2, p. 139–147 (Srovnání různých metod určení řádu autoregresního modelu při modelování průměrných měsíčních průtoků, in Czech).

[92] PROCHÁZKA M.: Log-Normal Distribution and the Possibility of Its Application to Hydrology. Hydrological Journal, **34**, 1986, No. 3, p. 243–256 (Logaritmicko-normální rozdělení a možnosti jeho použití v hydrologii, in Czech).

[93] QUENOUILLE, M. H.: Approximate Tests of Correlation in Time Series. J. Roy. Statist. Soc., B **11**, 1949, p. 68–84.

[94] RAIFFA, H.: Decision Analysis. Introductory Lectures on Choices under Uncertainty. Addison Wesley, Reading, Massachusetts – Menlo Park, California – London – Don Mills, Ontario 1970.

[95] REISENAUER, R.: Methods of Mathematical Statistics and Their Application. SNTL – Práce, Prague 1965, 210 p. (Metody matematické statistiky a jejich aplikace, in Czech).

[96] REZNIKOVSKIĬ, A. Š.: Hydroenergy Computations Using the Monte–Carlo Method. Energia, Moscow 1969, 296 p. (РЕЗНИКОВСКИЙ, А. Ш.: Водноэнергетические расчеты методом Монте–Карло, in Russian).

[97] ROGERS, P. P., FIERING, M. B.: Use of Systems Analysis in Water Management. Wat. Resour. Res., **22**, 1986, No. 9, p. 146S–158S.

[98] ROZHDESTVENSKYIĬ, A. V.: Estimation of the Precision of the Distribution Curves of Hydrological Characteristics. Gidrometeoizdat, Leningrad 1977, 270 p. (РОЖДЕСТВЕНСКИЙ, А. В.: Оценка точности кривых распределений гидрологических характеристик, in Russian).

[99] Directions for the Determination of Calculated Hydrological Characteristics. Gidrometeoizdat, Leningrad 1973, 110 p. (Руководство по определению расчетных гидрологических характеристик, in Russian).

[100] SALAS, J. D., DELLEUR, J. W., YEVJEVICH, V., LANE, W. L.: Applied Modelling of Hydrologic Time Series. Water Resources Publications, Colorado 1980, 484 p.

[101] SAVARENSKIĬ, A. D.: Methods of Computing Runoff Control, Gidrot. Stroit., 1940, No. 2 (САВАРЕНСКИЙ, А. Д.: Метод расчета регулирования стока, in Russian).

[102] Souček, V.: Analyses of the Relationships of Flows Series. Doctoral Dissertation, Prague 1965, 93 p. plus append. (Rozbory vztahů průtokových řad, in Czech).

[103] Souček V.: Analyses of the Relationships of Flows Series. Hydrological Journal, **13**, 1965, No. 1, p. 4–22 (Rozbory vztahů průtokových řad, in Czech).

[104] Souček, V., Vitha, O.: Computation of Long-term Reservoir-Controlled Runoff and Fluctuation of Solar Activity. Collection of Papers: Přehradní dny 1965, p. 77–93 (Výpočty víceletého regulování odtoku nádržemi a kolísání sluneční činnosti, in Czech).

[105] Sudler, C. E.: Storage Required for the Regulation of Stream Flow. Trans. of ASCE, Vol. 91, 1927.

[106] Svanidze, G. G.: Methods of Stochastic Modelling of Hydrological Series and Some Problems of Long-Term River Runoff Control. AN Gruz. SSR, 1961, Vol. 14, p. 189–216 (Сванидзе, Г. Г.: Методика стохастического моделирования гидрологических рядов и некоторые вопросы многолетнего регулирования речного стока, in Russian).

[107] Svanidze, G. G.: Fundamentals of Computation of River Runoff Control by Monte–Carlo Method. Izdat. Mecniereba, Tbilisi 1964, 272 p. (Сванидзе, Г. Г.: Основы расчета регулирования речного стока методом Монте–Карло, in Russian).

[108] Svanidze, G. G.: Mathematical Modelling of Hydrological Series. Gidrometeoizdat, Leningrad 1977, 296 p. (Сванидзе Г. Г.: Математическое моделирование гидрологических рядов, in Russian).

[109] Szolgay, J.: Stochastic Model of Daily Flows. Partial Research Report, Institute of Hydrology and Hydraulics of Slovak Academy of Sciences, Bratislava 1983, 45 p. (Stochastický model denných prietokov, in Slovak).

[110] Šor, J. B.: Statistical Methods of Analysis and Quality Control and Reliability. SNTL, Prague 1965, 456 p. (Statistické metody analýzy a kontroly jakosti a spolehlivosti, in Czech).

[111] Ventcelová J. S.: Probability Theory. ALFA/SNTL, Bratislava – Prague 1973, 524 p. (Teória pravdepodobnosti, in Slovak).

[112] Vitha, O.: The Effectiveness of Water Engineering Construction I–II. Doctoral Dissertation, Prague 1964 (Efektivnost vodohospodářské výstavby, in Czech).

[113] Vitha, O.: Some Notes on Long-term River Runoff Control. Collection of Papers: Přehradní dny, Jevany 1965 (Některé poznámky k víceletému regulování říčního odtoku, in Czech).

[114] Vorlíček, M., Holický M., Špačková, M.: Probability and Mathematical Statistics for Engineers. Czech Technical University, Prague 1982, 345 p. (Pravděpodobnost a matematická statistika pro inženýry, in Czech).

[115] Votruba, L. et al.: Analysis of Water Resource Systems. Elsevier, Amsterdam – Oxford – New York – Tokyo 1988, 454 p.

[116] Votruba, L., Brroža, V.: Reservoirs Water Management. Elsevier, Amsterdam – Oxford – New York – Tokyo 1989, 444 p.

[117] Votruba, L., Nacházel, K.: Fundamentals of the Theory of Stochastic Processes and Their Application to Water Engineering. Czech Technical University, Prague 1971, 183 p. (Základy teorie stochastických procesů a jejich aplikace ve vodním hospodářství, in Czech).

[118] White, J. B.: Probability Methods Applied to the Storage of Water in Impounding Reservoirs. Manchester University, Manchester, 1963.

[119] Yeh, W. W.-G.: Reservoir Management and Operations Models. A State-of - the-Art Review. Wat. Resour. Res., **21**, 1985, No. 12, p. 1797–1818.

[120] Yevjevich, V.: Fluctuations of Wet and Dry Years. Colorado State University, 1963–64.

[121] Yevjevich, V.: Stochastic Processes in Hydrology. Water resources publications, Fort Collins, Colorado, 1972.

Appendix of Bibliography

[122] International Conference on the Assessment of the Role of Carbon Dioxide and of Other Greenhouse Gases in Climate Variations and Associated Impacts. Villach, Austria, Oct. 1985, WMO No. 661.

[123] KLEMEŠ, V.: Geophysical Time Series and Climatic Change – a Sceptic's View. Lecture delivered at the Faculty of Civil Engineering of the Czech Technical University in Prague on April 10, 1990.

[124] KOS, Z.: Stochastic Water Requirements for Supplementary Irrigation in Water Resource Systems. IIASA, Laxenburg, Austria, RR–82–34, 61 p.

[125] KOS, Z.: Methods of Control of Irrigation from the Point of View of the Systems of Water Management. Partial Report of the National Plan of Basic Research II–5–7/6. Faculty of Civil Engineering of the Czech Technical University, Prague 1987, 45 p. and annexes (Metody řízení závlah z hlediska vodohospodářských soustav, in Czech).

[126] KOS, Z.: Mathematical Models of Water Management systems. Doctoral Dissertation. Faculty of Civil Engineering of the Czech Technical University, Prague 1989, 85 p. and anexes (Matematické modely vodohospodářských soustav, in Czech).

[127] RISSANEN, J.: Modelling by Short Data Description. Automatica, **14**, 1978, p. 465–471.

[128] SCHWARZ, G.: Estimating the Dimension of a Model. Ann. Stat., **6** 1978, p. 461–464.

[129] The Influence of Climate Change and Climatic Variability on the Hydrologic Regime and Water Resources. Proceedings of an International Symposium Held During the XIXth General Assembly of the International Union of Geodesy and Geophysics at Vancouver, Canada, August 1987, IAHS No. 168.

[130] The Changing Atmosphere: Implications for Global Security. Proceedings, Toronto, Canada, June 1988, WMO No. 710.

SUBJECT INDEX